WJTA Conference and Expo 2022

AA006182

New Orleans, Louisiana, USA
1 - 3 November 2022

ISBN: 978-1-7138-7626-7

Printed from e-media with permission by:

Curran Associates, Inc.
57 Morehouse Lane
Red Hook, NY 12571

Some format issues inherent in the e-media version may also appear in this print version.

Printed with permission by Curran Associates, Inc. (2023)

For permission requests, please contact WaterJet Technology Association-Industrial & Municipal Cleaning Association at the address below.

WaterJet Technology Association-Industrial & Municipal Cleaning Association
906 Olive Street – Suite 1200
St. Louis, MO 63101-1448
USA

Phone: (314) 241-1445
Fax: (314) 241-1449

www.wjta.org

Additional copies of this publication are available from:

Curran Associates, Inc.
57 Morehouse Lane
Red Hook, NY 12571 USA
Phone: 845-758-0400
Fax: 845-758-2633
Email: curran@proceedings.com
Web: www.proceedings.com

TABLE OF CONTENTS

2022 WJTA Conference & Expo
November 1-3, 2022 • New Orleans, Louisiana

Paper

AN EVALUATION OF RESPONSE TIMES FOR EXISTING AND POTENTIAL FUTURE PRESSURE RELEASE SAFETY DEVICES IN WATERBLASTING SYSTEMS

C. Hanley
StoneAge, Inc.
Durango, Colorado, U.S.A.

ABSTRACT

Pressure release devices are an integral part of waterblasting best practices and are mandated by contractors, asset owners and industry. These devices, commonly known as dump valves, are used as an operator's primary control to release pressure in the system, sometimes in response to a hazardous event. The design and response of these devices has a direct impact on the safety of persons in the vicinity of a waterblasting operation. The immediate response to a dangerous event not only depends on the operator's ability to react, but also on the equipment's response time and ability to reduce system pressure. The purpose of this research is to evaluate reaction times to an event by pairing operators with the most common pressure release devices including handheld pneumatic pilot operated valves, foot actuated pneumatic pilot operated valves and electronic pilot operated valves. The potential utility of automated responses will also be explored.

Organized and Sponsored by the WJTA®

1. INTRODUCTION

Pressure release devices, commonly referred to as dump valves in industry, are a critical component in waterblasting systems. The dump valve is responsible for giving the operator a quick and easy way to divert flow from the cleaning nozzle while also lowering the system pressure to a safe level. While pressure is not the only factor when classifying a waterjet to be in a safe state, skin simulants have shown to be penetrated by water jets at less than 500psi (Wright, 6).

A common style of dump valve is pneumatically piloted. A pneumatically piloted valve uses air pressure to shift the valve from one state to another, open to closed or vice versa. Tubing inner diameter (ID) and length are critical characteristics to consider when setting up a pneumatic circuit for use with a pressure release device. Different tubing ID's and lengths will lead to different valve response times. Generally, a larger volume of air in the pneumatic circuit will take more time to dissipate than a smaller volume. These dissipation times have a direct effect on the overall reaction time of the pressure release device. The dissipation time also depends on the initial pilot line pressure, commonly around 100psi, and the minimum pressure to shift a valve or allow a valve to spring return upon discharge.

An additional consideration is the energy stored in the high-pressure hoses. The larger ID hoses and lengths will release more energy in the event of a pressure release than a small ID and short hose for the same pressure.

Along with tubing ID, length, minimum pilot pressures and stored high pressure water energy, the means in which you initiate, or trigger, a change to the pilot signal is another variable of a pressure release device. Common triggers for pressure release devices are foot pedals, hand actuated levers or electronic control systems like Estop buttons or dead man levers. Each trigger design and implementation have pros and cons when considering reaction intuition, response time and reliability.

Lastly, the overall response time of a waterblasting system ultimately relies on the human reaction. The human reaction time has been a deeply researched field for different applications. For instance, a human automotive reaction-time study reported an average reaction time of 0.627 seconds (Nagler, 261). To determine reaction times in a waterblasting environment, operators of varying experience levels were paired with different dump valve triggers to perform a reaction-time study of a waterblasting event. In addition, the benefits of using computerized reactions to the same event were explored.

2. TEST ARRANGEMENT

2.1 Pneumatic Pilot Pressure

2.1.1 Threshold Pilot Pressure Determination

This testing was designed to determine at what pneumatic pilot pressure a standard 20kpsi dump valve would shift from a closed state, sending water to the nozzle, to and open state, allowing water to divert from the nozzle and lower the system pressure to a safe level. To determine this, the pneumatic pilot pressure was adjusted using a regulator and measured using a ProSense 0-200psig SPTD25-20-0200H pressure transducer. With the valve shifted to the closed state, the water pressure was brought up to 20kpsi, using a 325hp diesel pump and measured using a Core Sensors 0-44kpsi CS10-2A44000PS5A000-03 pressure transducer. The pneumatic pilot pressure was then slowly decreased to zero using the regulator. The data from the water pressure transducer and pneumatic pilot pressure transducer could then be correlated to determine the pilot threshold pressure of the dump valve.

2.1.2 Discharge Time

This testing consisted of pairing different tubing diameters and lengths with an industry standard dump valve. The dump valve used was a spring return style piloted by a pneumatic cylinder. Air pressure in the pilot signal was measured using a ProSense 0-200psig SPTD25-20-0200H pressure transducer. Discharge times from 100psi to the threshold pressure described above in section 2.1.1 Threshold Pressure Determination were recorded for each ID and length combination. The tubing sizes tested had IDs of .125", 156" and .250". Each tubing size was tested at lengths of 5', 10', 15', 20', 50' and 100'. All tubing ID and length combinations were discharged and recorded three times.

2.2 Water Discharge Impact and Pressure

To evaluate the energy stored in different high-pressure systems, conditions on the bypass or diversion outlet of an industry standard dump valve were measured. To do this, the bypass outlet of the dump valve, normally a free flowing, large diameter orifice, was restricted using a .165" inner diameter nozzle. Pressure just upstream of the nozzle was measured using a Core Sensors 0-44kpsi CS10-2A44000PS5A000-03 pressure transducer. In addition, the discharge impact force was evaluated by orienting the nozzle normal to a target at an approximate 2" standoff. Impact force on the target was measured using an Omega 0 -1000lb LC1103-1K load cell. 50' lengths of high-pressure hose ID's of 3mm, 5mm, 8mm and 13mm were tested at operating pressures of 5, 10, 15 and 20kpsi. The bypass event was initiated by manually releasing a pneumatic foot pedal trigger. All hose ID and pressure combinations were discharged and recorded a minimum of five times.

2.3 Human Reaction

To simulate a safety concerning event in which the operator felt a need to react by using the dump valve, rupture disc failures were used. The idea behind using a rupture disc bursting is that it closely emulates visual and auditory characteristics of a safety concerning failure such as fitting failures or the nozzle coming off the end of a high-pressure hose.

Unbeknownst of the startling rupture disc event, operators of different skill levels were asked to clean a heat exchanger with automated equipment. This is shown in the left image of Figure 5. While the operator was focused on positioning and feeding the high-pressure hose, the pump pressure was slowly increased above the rupture disc rating. The rupture event is shown in the right image of Figure 5. The different automated equipment and dump valve trigger configurations used in this testing are shown below.

Automation Control	Trigger for Dump Valve	Automated Equipment
StoneAge CBX Control Box	Control Box Lever	StoneAge Compass
StoneAge CBX Control Box	Foot Pedal	StoneAge Compass

StoneAge CTRL-100 with AutoMove* and AutoFeed*	Automation Lever or Estop	StoneAge Compass with Sentinel Technology
StoneAge CTRL-100 with Mapping*	Automation Lever or Estop	StoneAge Compass with Sentinel Technology

*AutoMove – Automated positioning to the next tube location with just a toggle of the CTRL-100 joystick.
*AutoFeed – Automated feed in and feed out with just one click of the CTRL-100 forward feed button.
*Mapping – One click of the X/Y positioner joystick on the CTRL-100 to clean the entire exchanger. User simply acts in an overseer capacity.

Layouts of the controls systems for the CBX Control Box and CTRL-100 can be seen in the left and right images of Figure 6, respectively.

2.4 Computer Reaction

Based on the consistent water pressure trends found during a rupture event, automated computer reactions were tested against the rupture disc failure condition. Similar to the human reaction testing, water pressure was slowly increased above the pressure rating of the rupture disc. Timing of the computer response was captured by comparing the falling edges of the water pressure and pneumatic pilot pressure. The computer reacted by changing the state of a relay which the Estop signal passed through.

3. DATA ANALYSIS

3.1 Pneumatic Pilot Pressure

3.1.1 Threshold Pilot Pressure Determination

To correlate the pneumatic pilot pressure to water pressure, timestamps of each data set were compared to synchronize the data points shown in Figure 1. The threshold pressure was determined by the pilot pressure which corresponded to a water pressure less than 500psi.

3.1.2 Discharge Time

The discharge time for each tubing ID and length pair was determined by averaging three discharges events. The discharge time is defined by the time it takes to go from the initial pilot pressure, 100psi, to a pilot pressure less than the threshold pilot pressure of the system defined in 4.1.1 Threshold Pilot Pressure Determination. More specifically, these values are the difference between the falling edge of the pilot pressure data to the first data point less than the threshold pressure. This time window is represented between the two dashed vertical lines in Figure 2.

3.2 Water Discharge Impact and Pressure

The impact force and pressure for each high-pressure hose ID and operating pressure pairing was determined by finding the maximum force and pressure that occurred over a minimum of five discharge events per high-pressure hose ID and operating pressure combination.

3.3 Human Reaction

To determine how long it took each person to react to the rupture disc event, timestamps of the water pressure data and pneumatic pilot pressure data from the dump valve circuit were compared.

Timing of the rupture event could be determined by the falling edge of the water pressure data and a reaction event could be determined by the falling edge of the pneumatic pilot pressure. An example of this reaction time determination is shown in Figure 7 by the spacing between the two dashed vertical lines.

3.4 Computer Reaction

With the data compiled from the 18 human reaction tests, it was determined that the slope of the decaying water pressure due to a rupture event is a condition that could be used to trigger a reaction by the computer. The 18 rupture events produced an average water pressure decrease of 44kpsi per second with a standard deviation of 6.83kpsi. Using an exclusion of two standard deviations the threshold for a reaction trigger was set to a water pressure decrease of 30.34kpsi per second. With an average sample rate of .026 seconds, the computer performed a rolling least-squares linear regression on the last N data points in time. If the slope of the linear regression dropped below -30.34kpsi per second, the computer changed the state of the relay to trip an Estop condition. Determining the computer reaction times used the same method outlined above in 3.3 Human Reaction and can be seen in Figure 8.

4. TEST RESULTS

4.1 Pneumatic Pilot Pressure

4.1.1 Threshold Pilot Pressure Determination

Shown by the data in Figure 1, the valve does not have binary states of on and off. Due to the characteristics of a spring return valve assembly, there exists a pilot pressure range between states of flow being directed to the nozzle or directed to the bypass or diverted state. Between the approximate pressures of 7-13psi, some of the water is going to the nozzle and some is being diverted, still maintaining a certain amount of system pressure at the tool. Based on the safe operating pressure of 500psi mentioned above, the threshold pilot pressure of this system was determined to be any pressure less than 7psi.

4.1.2 Discharge Time

A consistent trend across all pneumatic pilot circuits is that discharge time increases with volume. With the initial pressure and size of the pressure release orifice held constant, the larger the air volume present in the tubing, the longer it will take to discharge to atmosphere. However, if there exists a long length of tubing combined with a proportionally small inner diameter, the effects of pressure loss could start to dominate the discharge timing. This is represented in Figure 3 where the 100’ of .125” ID tubing had a longer discharge time than the 100’ of .250” ID tubing.

4.2 Water Discharge Impact and Pressure

As demonstrated in this test, both impact force and discharge pressure increase with system pressure for any given hose ID, as shown in the left and right images of Figure 4, respectively. In addition, the ID of your high-pressure hose has a significant impact on the potential energy stored in the high-pressure system. As the high-pressure hose ID increases, so does the potential energy in that system. This is supported by the increase in impact force and discharge pressure for the larger ID’s relative to the smaller ones at a given pressure.

4.3 Human Reaction

The 18 human reaction tests produced an average response time of 1.14 seconds with a standard deviation of 0.83 seconds.

4.4 Computer Reaction

Using a rolling sample of five data points, the computer produced an average reaction time of 0.11 seconds with a standard deviation of 0.04 seconds across six tests. Changing to a rolling sample size of three data points produced an average response time of 0.07 seconds with a standard deviation of 0.01 seconds across four tests.

5. CONCLUSIONS

- When possible, trending towards smaller tubing ID's and lengths in pressure release pilot circuits will help maintain a reasonable dump valve response time. If small tubing ID's and lengths are not feasible due to job characteristics, consider having all persons a further distance away from any high-pressure hoses or connections. Results of this paper in conjunction with EN ISO 13855 can be used as references for reasonable operator standoff distances.
- When working with larger volumes of high-pressure water (i.e., larger ID hoses and lengths or multiple hoses) it is recommended that the pressure relief valve be rigidly mounted, and all connections be tethered by whip checks. As the volume of high-pressure water increases, so does the energy released by that system during a failure or pressure release event. Lastly, the orientation of the dump valve bypass or diversion orifice should be considered as dangerous water velocities and volumes can be released here.
- Pneumatic foot pedals and control boxes are still great options as they are intuitive for users.
- Electronic triggers offer an avenue for future growth. Computerized reactions to sudden drops in system pressure can be used to respond to a safety concerning waterblasting event quicker and more consistently than a human.

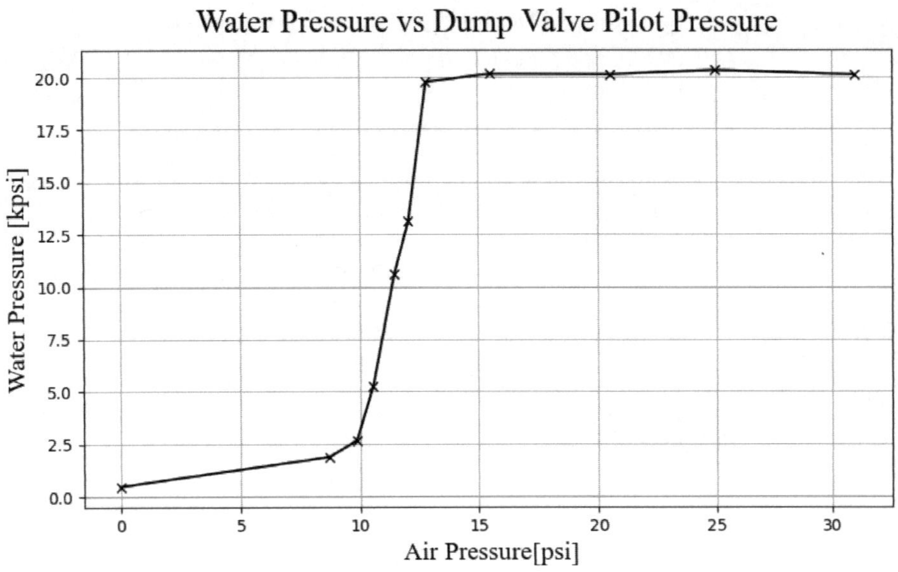

Figure 1

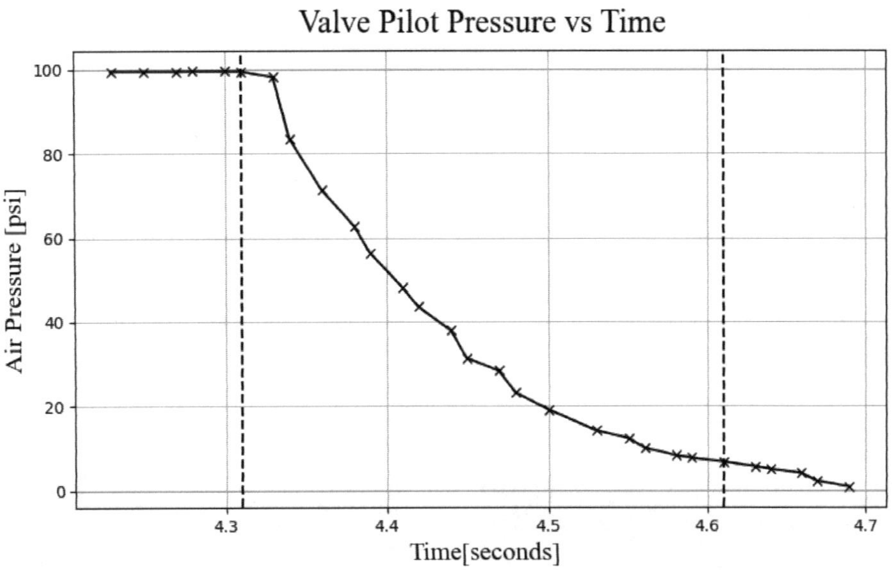

Figure 2

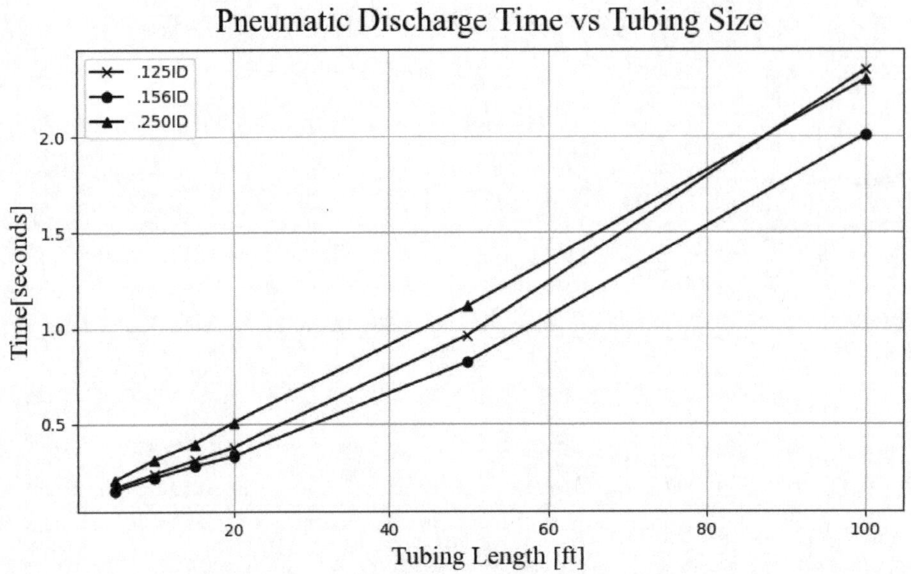

Figure 3

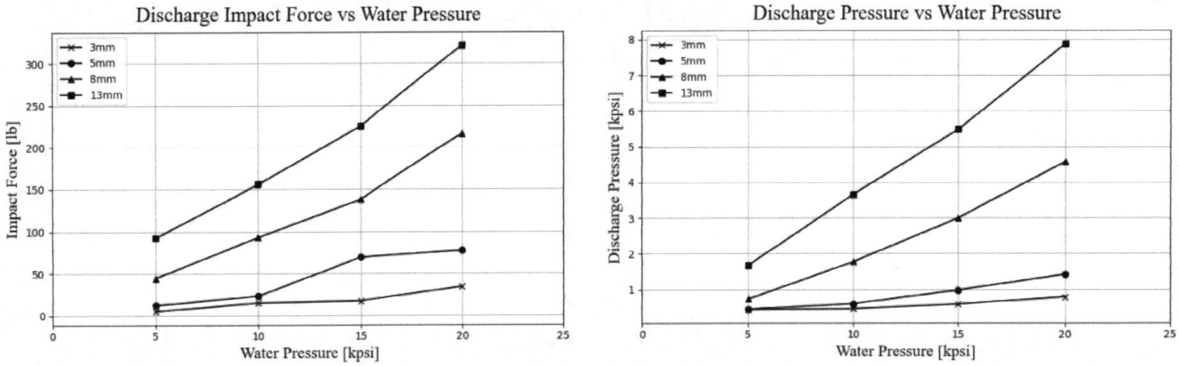

Figure 4

Normal Operation Rupture Event

Figure 5

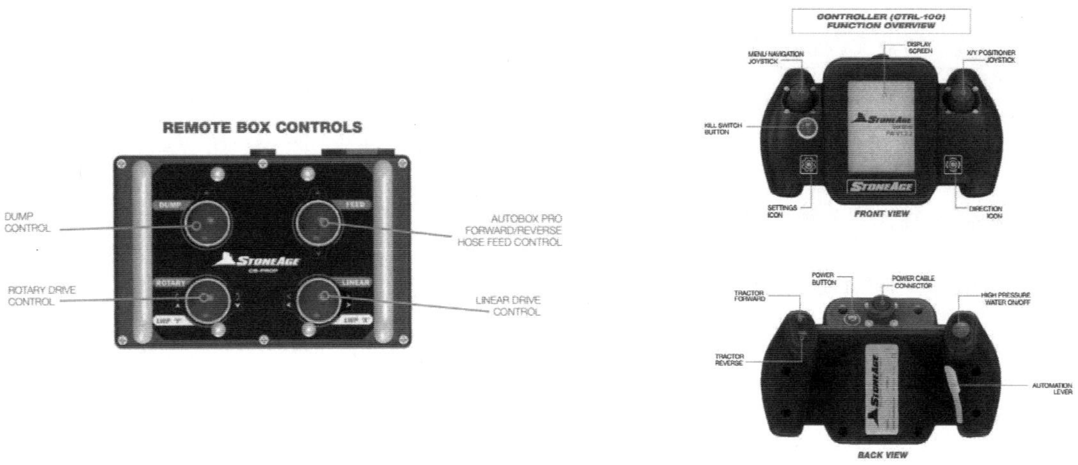

Figure 6

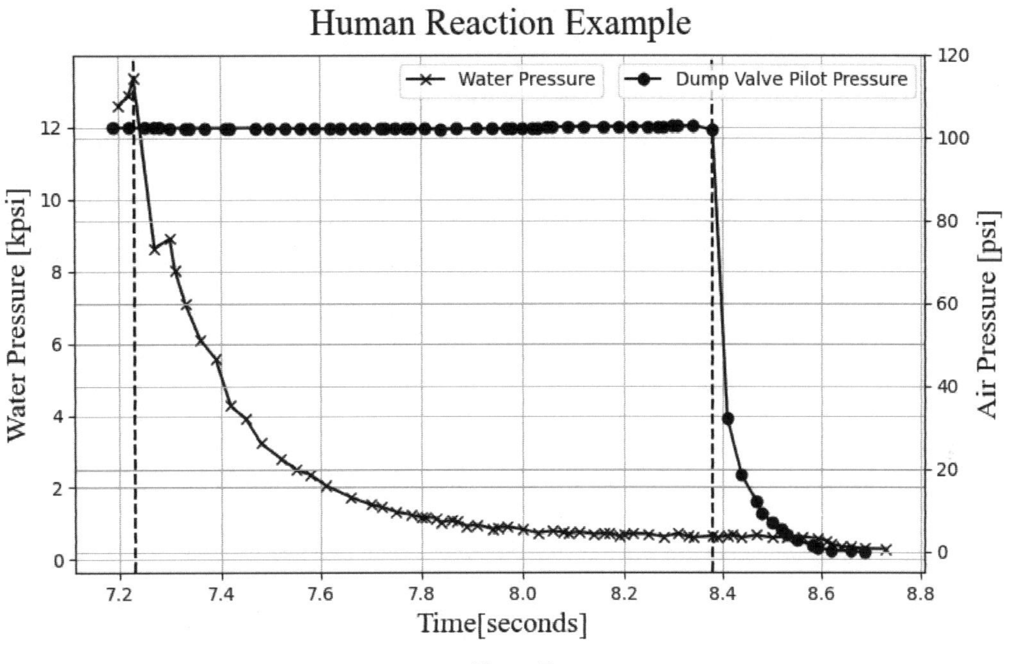

Figure 7

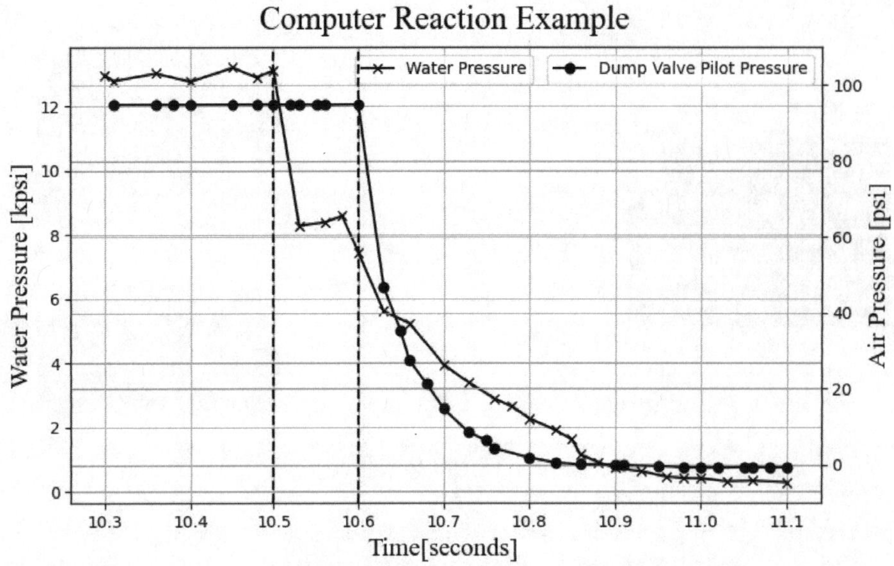

Figure 8

6. REFERENCES

Nagler, Charles. Nagler, William. "Reaction Time Measurements". *Forensic Science*, vol. 2, 1973, pp. 261-274.

Wright, Doug. "Safety Considerations for Low Pressure Cleaning Applications." *WJTA American Waterjet Conference*, November 11-13, 2019, New Orleans, Louisiana, 2019

2022 WJTA Conference & Expo
New Orleans, Louisiana

Paper

HEAT EXCHANGER TUBE MATERIAL RESPONSE TO HIGH PRESSURE WATERJET IMPACT AND THE FACTORS AFFECTING THE RISK OF DAMAGE DURING THE CLEANING PROCESS

D. Wright
StoneAge, Inc.
Durango, Colorado, U.S.A.

ABSTRACT

The utilization of waterjets for the cleaning of heat exchanger tubes is widely accepted in industries around the world. The fouling or scales that form during use typically determine the specified operating pressure of the waterjet cleaning system, often in the range of 69 to 276 MPa (10,000 to 40,000 psi). While carbon steel and stainless steel tube materials are reasonably resistant to properly applied waterjet cleaning techniques, other softer metallic materials consisting of brass, nickel, and copper alloys are commonly used in tube and shell heat exchanger construction and their potential for damage by high pressure jet erosion is expected to be greater. While operating pressure is one of the primary factors in the risk of damage to all tube materials, other parameters of the cleaning system play a role, such as the movement of the jets by rotation of the lance or nozzle head, the rate or pause of translation of the nozzle head through the tube, and the jet angle of impingement on the inside of the tube wall. The purpose of this research was to test and present the effect of increasing jet pressure combined with other parameters that determine the probability of the erosion that may occur on common metallic tube materials such as carbon and stainless steel, brass, copper, titanium, and nickel alloys.

Organized and Sponsored by the WaterJet Technology Association

1. INTRODUCTION

The impact of a plain (not abrasive) waterjet on a metallic surface can remove material through the process of cavitation erosion. The amount or depth of material removal is primarily dependent on the material type, the pressure of the jet, if the jet is fixed or in a rotating nozzle head, and the time that the jet is left in one place without being continuously translated, whether rotating or not. Standoff distance of the jet to the tube wall, jet angle of impact and orifice size also affect the rate of material removal by lesser amounts.

The selection of operating pressure for the cleaning of heat exchanger tubes is generally based on operator experience in removing specific types of deposits from the tubes and does not always take into consideration the tube material, nozzle type, or the methods of the workers performing the cleaning. As an example, it is a common practice to place the nozzle in a stationary position inside a clean tube while setting the operating pressure. A worker might also leave a nozzle in place against a hard deposit or plug in the tube, waiting for the jets to clear the material instead of continuously moving the nozzle. The potential amount of damage in either case is also dependent on whether the nozzle is rotating or stationary, as well as the time that the nozzle is left in one place. Alternatively, a trained and aware worker or an automated system programmed to not allow a nozzle to be left in one place while at pressure might allow acceptance of higher operating pressures for cleaning if needed for faster or more effective cleaning.

This paper presents the potential of cavitation erosion with the variables of pressure, time of exposure, static jets and rotating jet heads on different materials commonly used in heat exchanger construction, to allow owners of equipment and waterblast cleaning services to determine risk of damage, and to specify cleaning pressures combined with methods or controls of application to optimize the cleaning process while minimizing the potential risk.

2. TEST PROCEDURES

The material samples used in the tests consisted of tube and flat plate, depending on material availability. The plate samples were the primary means as they were simpler to set up and measure after tests were performed. Table 1 lists the materials and designations.

The first series of non-rotary tests performed on each material type consisted of a typical nozzle head with radial jets installed on a fixed piece of high pressure tubing with one jet aimed directly at the surface, set at a standoff distance of 4.8 mm from the material surface. Table 2 describes the nozzle heads used in the tests. The system was brought to the desired test pressure and the head traversed onto the material and paused for each separate time of exposure. Figure 1 illustrates this nozzle head and test arrangement.

The second series of tests was performed with the same nozzle head type installed on a self-rotating tube nozzle assembly, with a standoff distance of 4.8 mm from the material surface except when determining effect of varied standoff distance. The system was again brought to the desired test pressure and the spinning head traversed onto the material and paused for each separate time of exposure. Figure 2 illustrates this tool and test arrangement. The effects of larger and smaller orifice sizes were also compared with the rotary test.

The third series of tests consisted of traversing the rotating nozzle head back and forth across an area of the sample material with up to 100 passes at a rate of 50 mm/s to evaluate the effect of repeated cleaning cycles with a continuously moving and rotating nozzle.

3. TEST RESULTS

3.1 Non-Rotating Jet Effect

The results of the non-rotary jet testing for all the materials are shown in Figure 3, where a test pressure of 103 MPa (15,000 psi) was applied through Nozzle Head "A" with an orifice size of .9 mm (.036 in) impacting the material, for time periods of 10, 30, and 60 seconds. The soft copper and copper alloys showed the greatest effect, while the Monel, Titanium and Stainless Steel show the least. This chart illustrates the relative difference of effect at the same pressure between all the material types as a reference point. For all materials, the non-rotating jet did not remove any material when traversed between the timed test points.

From these results, the materials could be divided into "Soft" and "Hard" as to their resistance to erosion by jet impact. Figure 4 plots the results of the non-rotary jet on Copper, Copper-Nickel 90/10, and Brass 260 at the lower pressures of 55 MPa (8,000 psi) and 69 MPa (10,000 psi). If no controls on operational methods were in place, the pressure might need to be limited still further for the Copper and Copper-Nickel alloys to avoid damage to the tube wall.

The relatively "Hard" materials, with greater resistance to non-rotary jet erosion were the Monel, 1018 CS, 304 SS and Titanium, with the 304 SS being most resistant, followed by the Titanium. Figure 5 shows the results in these materials with the non-rotary jet impact at pressures of 69 MPa (10,000 psi), 103 MPa (15,000 psi) and 138 MPa (20,000 psi). Without controls on operation, maximum pressures for these materials might need to be limited as well.

3.2 Rotating Jet Effect

Rotating the nozzle head results in continuous motion of the jet across the surface, which by itself reduces the amount of potential damage to the surface, although if left rotating in the same place for periods of time, erosion can still occur. The application of the rotating nozzle head allows potentially higher operating pressures for any material. Figure 6 illustrates the difference between non-rotary and rotary at 103 MPa (15,000 psi) for the Copper-Nickel 90/10 and 1018 Carbon Steel; the depth of material removed can differ by a factor of 10 for the same pressure and time of exposure.

From Figure 7, a rotary head pressure of 69 MPa (10,000 psi) could be acceptable for the Copper-Nickel 90/10 but still at risk for the Copper 110 material without operational controls. Figure 8 shows the resulting effect when the pressure increased to 103 MPa (15,000 psi) in these same materials as well as 1018 Carbon Steel; it would be necessary to prevent the nozzle head from stopping translation to avoid tube wall damage at this pressure in these materials.

Figures 9 and 10 show the results of rotary head testing at 103 MPa (15,000 psi) and 138 MPa (20,000 psi) on Monel 400, 304 Stainless Steel, 1018 Carbon Steel and Titanium. The 304 SS was the most resistant, followed by Titanium and Monel, which essentially fall in between the 304 SS and the 1018 Carbon Steel in their resistance to cavitation erosion. Figure 11 shows the result in the same materials at 228 MPa (33,000 psi) with reduced times of exposure to demonstrate the importance of not allowing the nozzle head translation to be paused during the cleaning process with higher operating pressures.

3.3 Standoff Distance, Jet Power and Jet Angle Effects

The effect of standoff distance was tested with the rotating nozzle head arrangement within the range that might be found in tube cleaning, and as has been previously shown with cavitation, there is an optimum distance for the greatest effect, and no effect if the jet orifice is against the surface. This response is related to how the jet breaks up into droplets as it travels through the air or water before contacting the surface. Figure 12 shows the results of these tests conducted on 304 Stainless Steel at 138 MPa (20,000 psi) and Copper-Nickel 90/10 at 69 MPa (10,000 psi) with an exposure time of 30 seconds at each point.

The effect of increasing the jet power through increased flow rate was also tested with the rotating nozzle head arrangement, in 1018 Carbon Steel at 138 MPa (20,000 psi), comparing the Nozzle Head "A" with a .91 mm (.036") orifice size to Nozzle Head "D" with a .66 mm (.026") orifice size. At short exposure time periods there was minimal difference, while at the maximum 60 second exposure the higher flow rate head did remove more material. These results are shown in Figure 13.

The effect of jet angle of impingement on the surface of the material was tested to determine if there would be a reduction in damage with less direct impact. This did prove to be true at a 15 degree angle of impact, but at a 30 degree angle to the surface, the damage had increased to nearly half that of the 90 degree impact result, as shown in Figure 14, conducted on the Copper 110 plate at 138 MPa (20,000 psi) for 60 seconds.

3.4 Rotating and Traversing Jet Effects

This series of tests was conducted by traversing the rotating nozzle arrangement back and forth for multiple passes across the same surface area, at a rate of 50 mm/s. None of the tests produced a measurable erosion rate, although the surface generally ended up with a frosted appearance. This highlights the importance of maintaining continuous motion to minimize or avoid damage to the tube wall surface in any material. Figure 15 shows Copper 110 after test at 103 MPa (15,000 psi) and 100 passes across the surface. Figure 16 shows 1018 Carbon Steel after test at 228 MPa (33,000 psi) and 50 passes; Figures 17 and 18 show Monel and Titanium after 50 passes each at 228 MPa (33,000 psi).

4. CONCLUSIONS

This testing was performed to aid in the specification of heat exchanger tube cleaning processes, from operating pressure to types of nozzle heads and personnel training for operating procedures. There exists a pressure threshold below which would result in minimal damage to the tube wall over an extended period of time of jet exposure; however, this pressure threshold may be below that needed to effectively remove a deposit at a reasonable rate. Utilizing training or automated controls to ensure that the nozzle head maintains rotation and feed rate while operating at cleaning pressure is the most effective means of avoiding damage to the tube wall.

Regarding the material types and their resistance to jet erosion, it was found that the Copper and Copper Nickel alloys would be most susceptible to damage with higher pressures, and operating pressures should be limited in the range of 48 to 69 MPa (8,000 to 10,000 psi) at maximum. The Monel 400, Titanium Type 2 and 304 Stainless Steel exhibited the greatest resistance to damage by jet erosion, and operating pressures in the range of 103 to 138 MPa (15,000 to 20,000 psi) with moderate operational control would be acceptable. Higher cleaning pressures in the range of 200 to 280 MPa (30,000 to 40,000 psi) should always have operational controls in place to limit or avoid jet damage.

Material	UNS	Form	Other Common Names
Copper-Nickel 90/10	C70600	Sheet	Cupro-Nickel
110 Copper	C11000	Sheet	CDA 110
101 Copper	C10100	Tube	C101
260 Brass	C26000	Tube	Cartridge Brass
400 Nickel	N04400	Sheet	Monel 400
304 Stainless Steel	S30400	Plate	304 Stainless Steel
1018 CDF Carbon Steel	G10180	Plate	1018 Carbon Steel
Titanium Grade 2	R50400	Sheet	Titanium Type 2

Materials and Designations Used in Testing
Table 1.

Head Name	Orifice Type	Pattern 1	Orifice Size	Pattern 2	Orifice Size
"A"	Drilled	2 X 90°	.91 mm (.036")	2 X 105°	.81 mm (.032")
"D"	Drilled	2 X 90°	.66 mm (.026")	2 X 105°	.66 mm (.026")
"UHP"	Sapphire	2 X 85°	.25 mm (.010")	2 X 110°	.36 mm (.014")

Nozzle Heads Used in Testing
Table 2.

Non-Rotary Head Test Arrangement Shown with Copper Nickel 90/10 Sheet
Figure 1.

Self-Rotary Tube Nozzle Assembly Test Arrangement and Results in Copper Nickel
Figure 2.

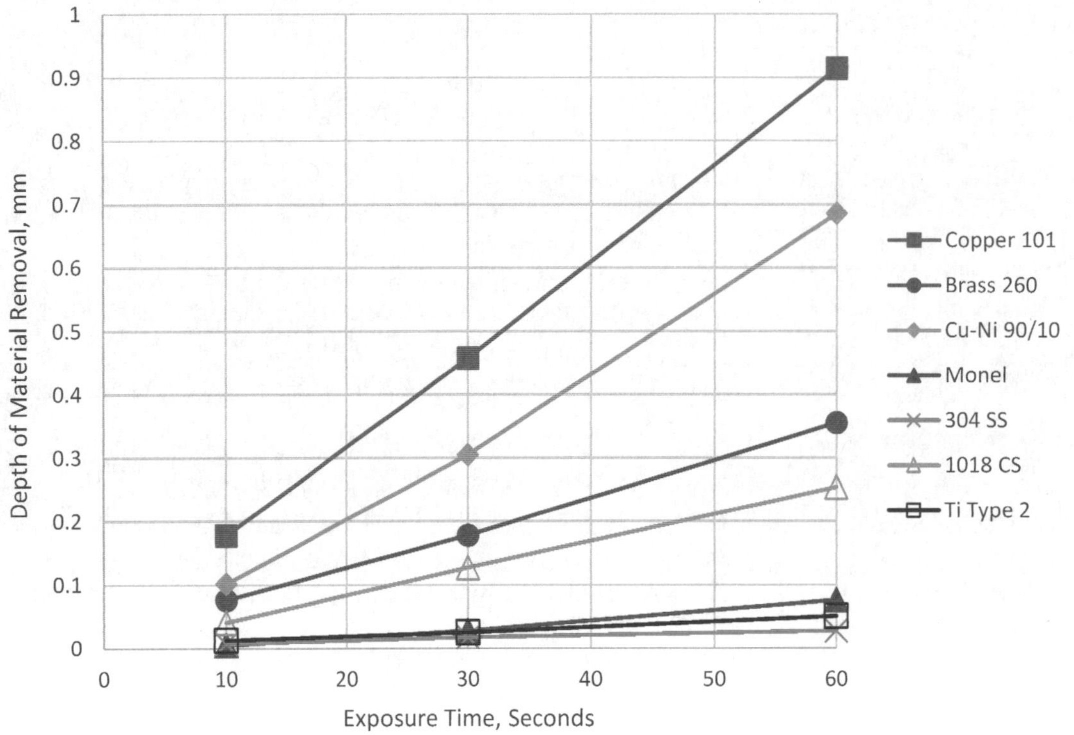

Relative Effect of Non-Rotary Jet Applied at 103 MPa (15,000 psi)
Figure 3.

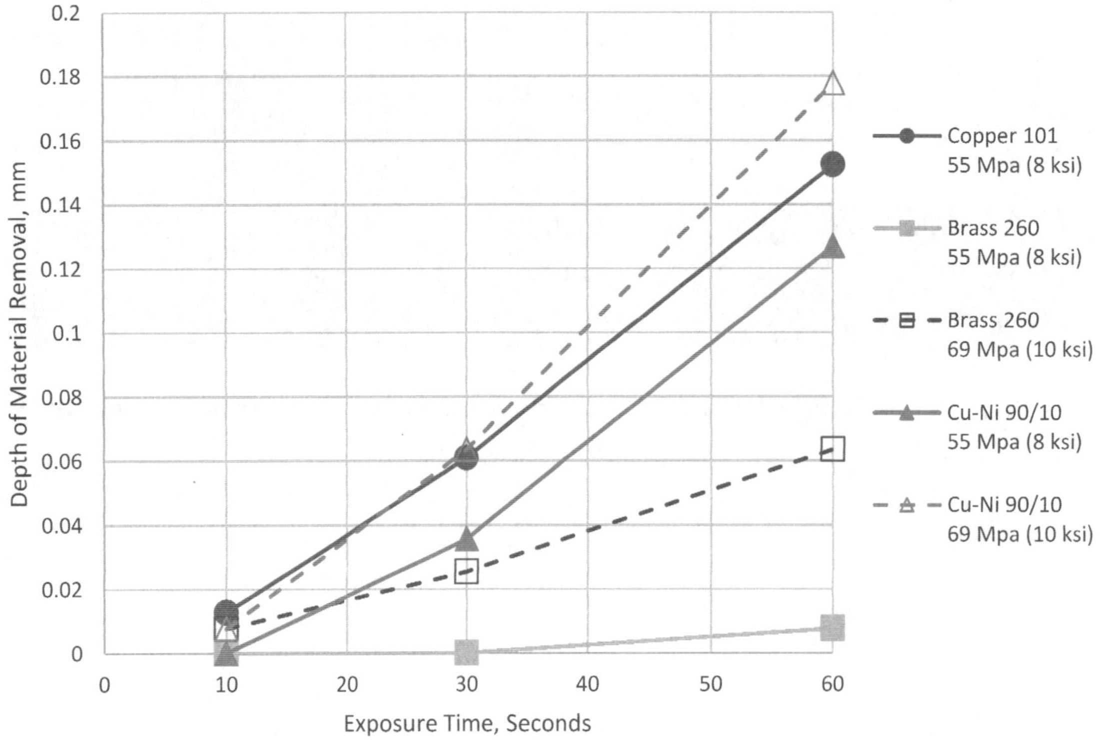

Effect of Non-Rotary Jet at Lower Pressures for "Soft" Materials
Figure 4.

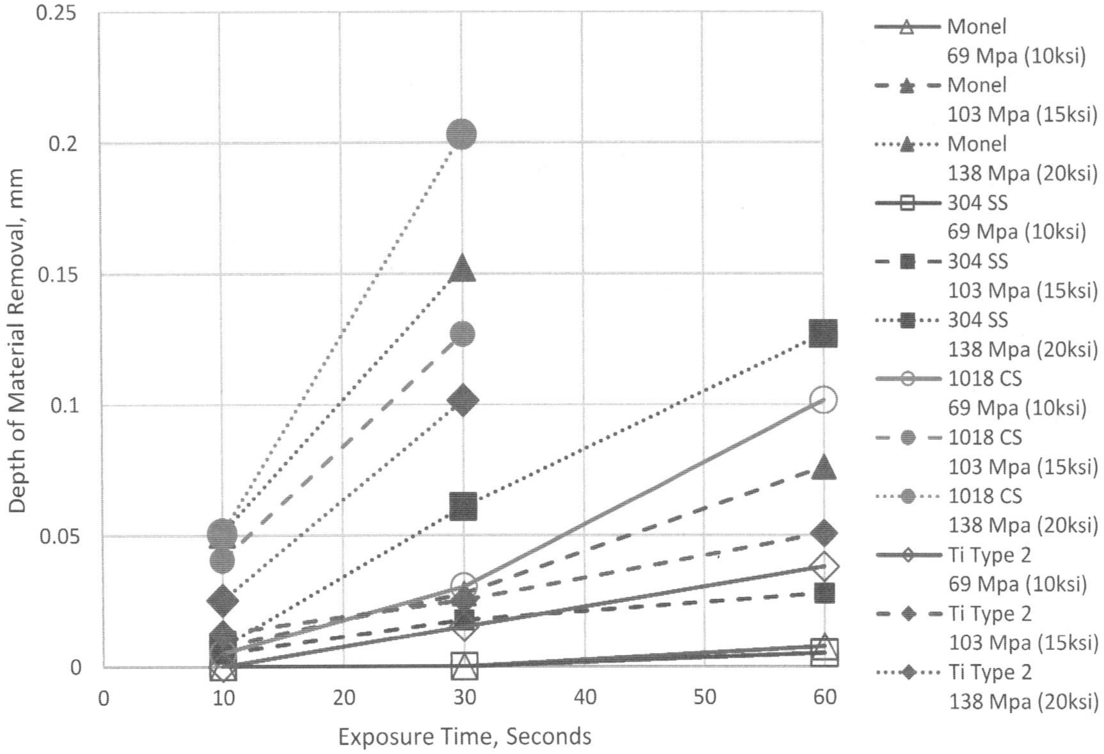

Effect of Non-Rotary Jet at Pressures up to 138 MPa (20,000 psi) for "Hard" Materials
Figure 5.

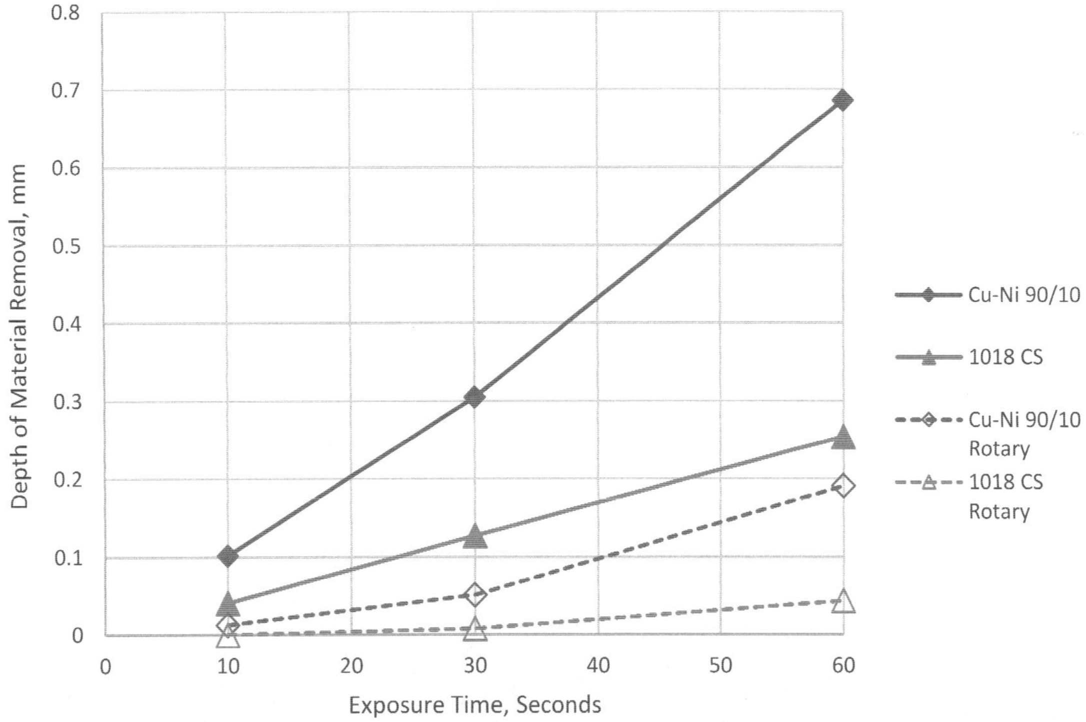

Comparison of Rotating Head to Non-Rotary in Two Materials at 103 MPa (15,000 psi)
Figure 6.

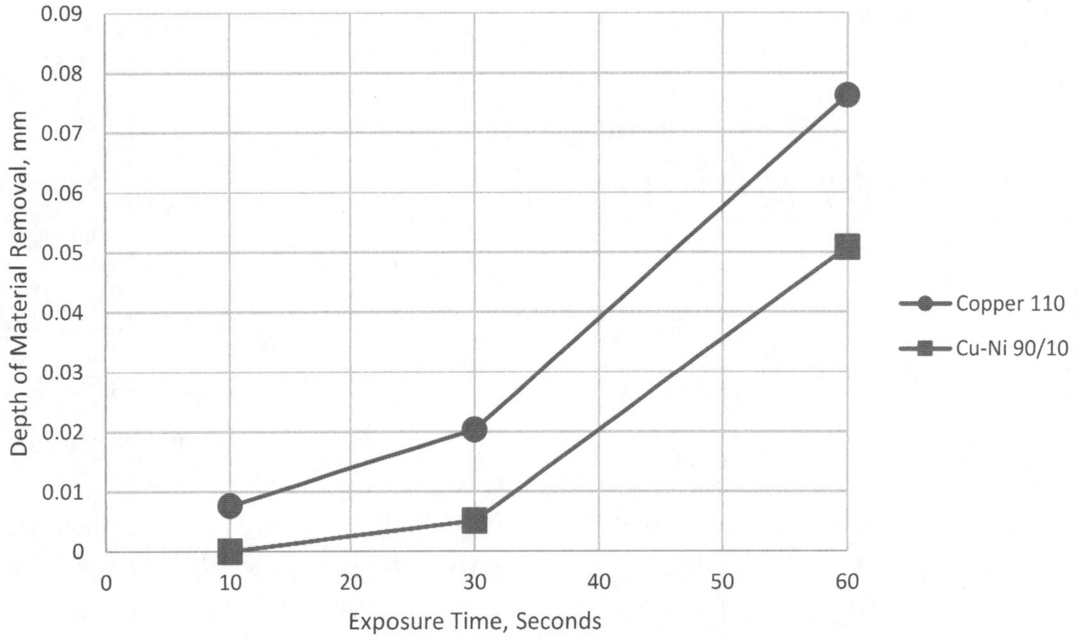

Copper 110 and Copper-Nickel 90/10 with Rotating Head at 69 MPa (10,000 psi)
Figure 7.

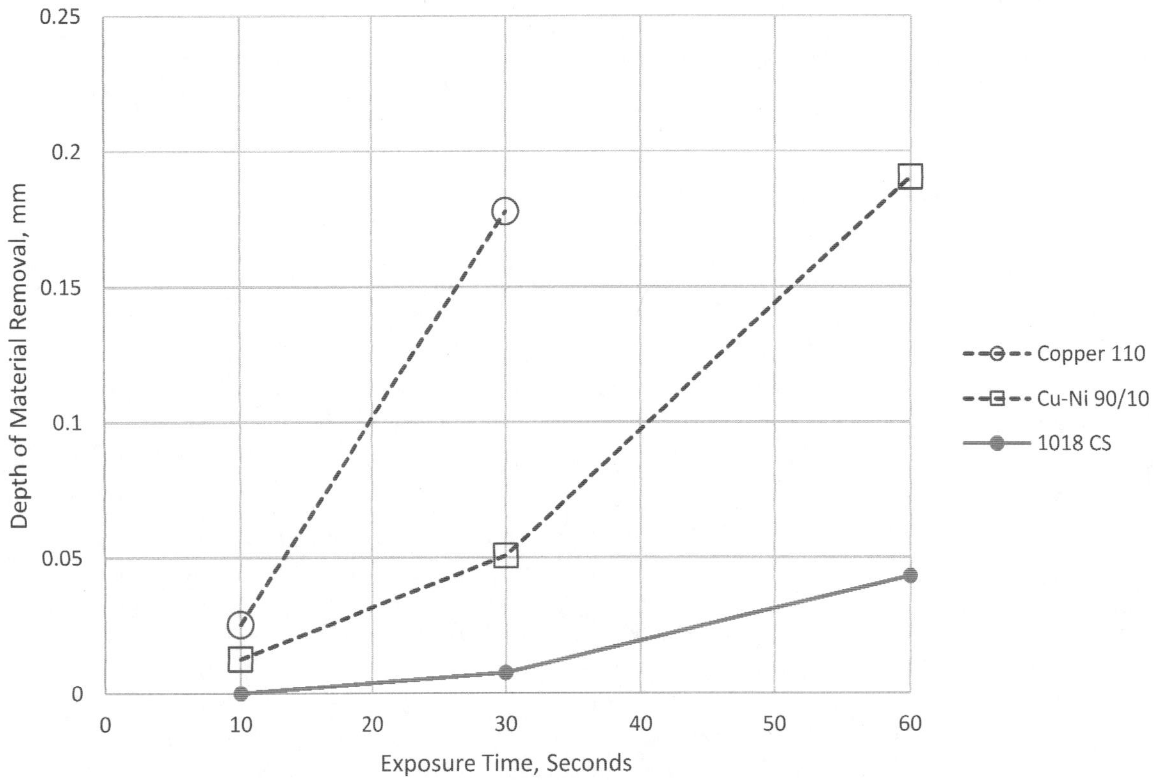

Copper 110, Copper-Nickel 90/10 and 1018 CS with Rotating Head at 103 MPa (15,000 psi)
Figure 8.

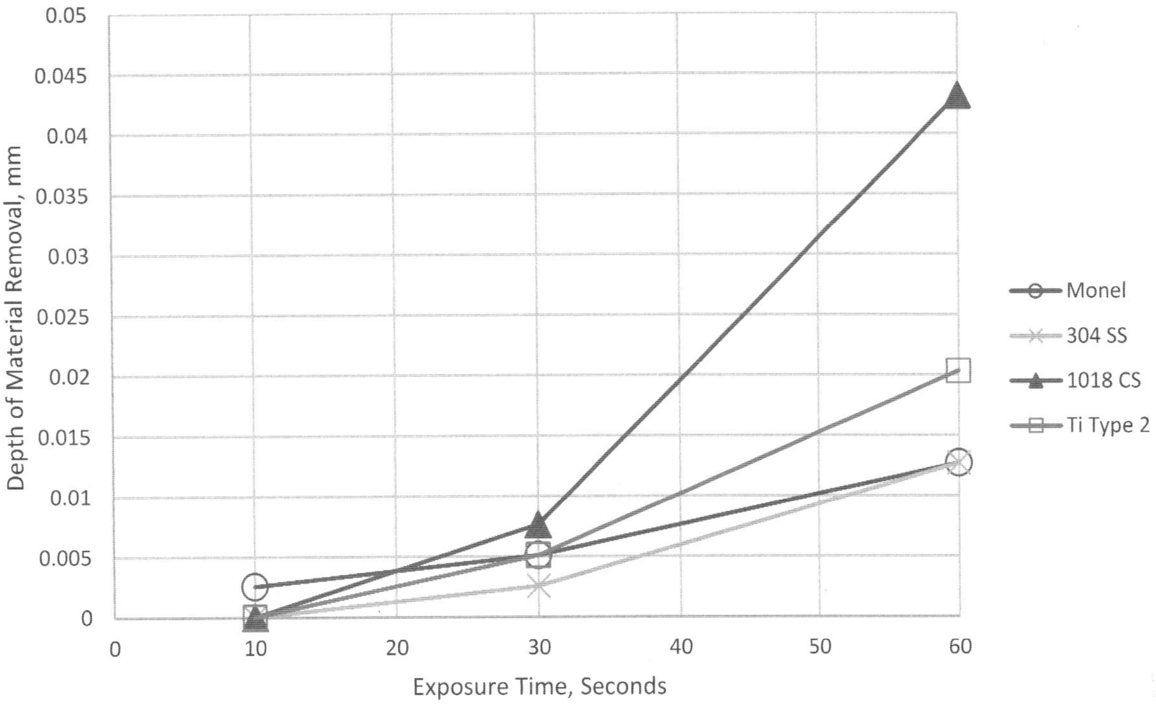

Monel, 304 SS, 1018 CS and Titanium with Rotating Head at 103 MPa (15,000 psi)
Figure 9.

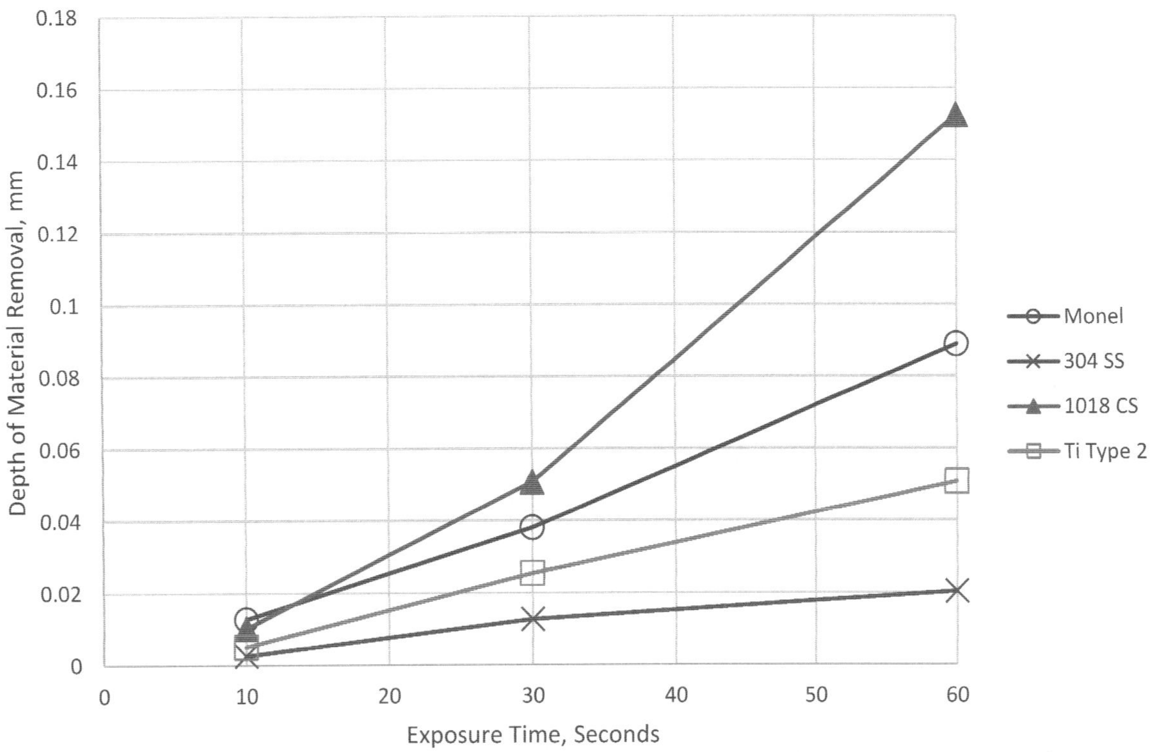

Monel, 304 SS, 1018 CS and Titanium with Rotating Head at 138 MPa (20,000 psi)
Figure 10.

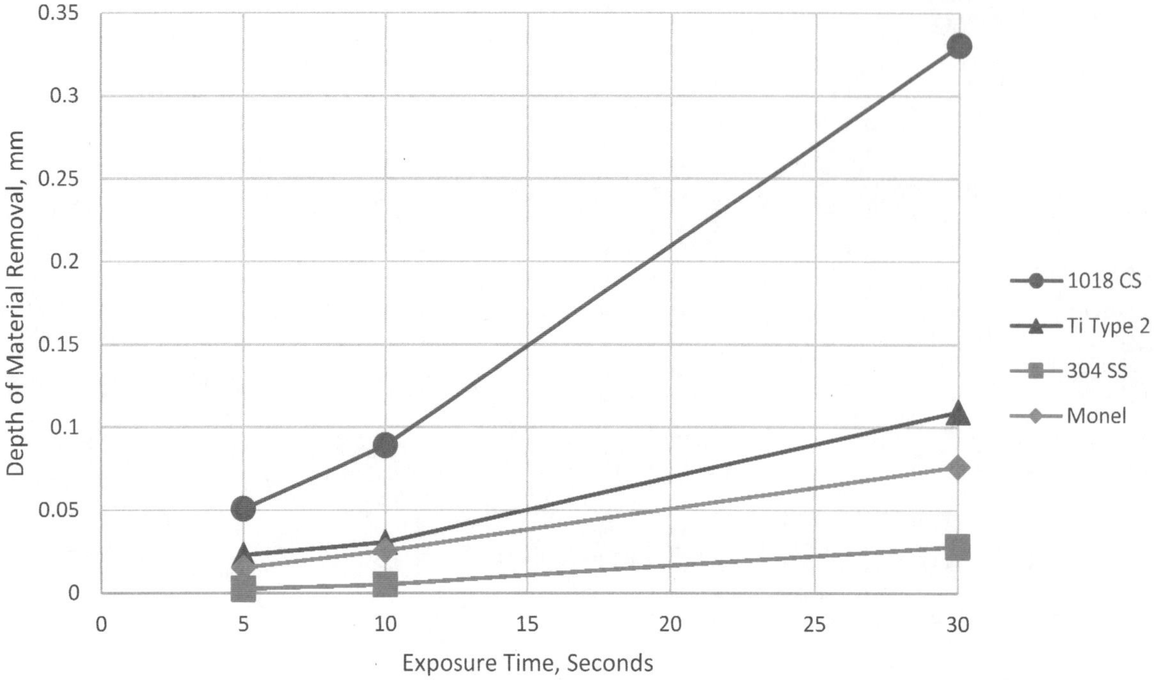

Monel, 304 SS, 1018 CS and Titanium with Rotating Head at 228 MPa (33,000 psi) at Exposure Times of 5, 10 and 30 Seconds
Figure 11.

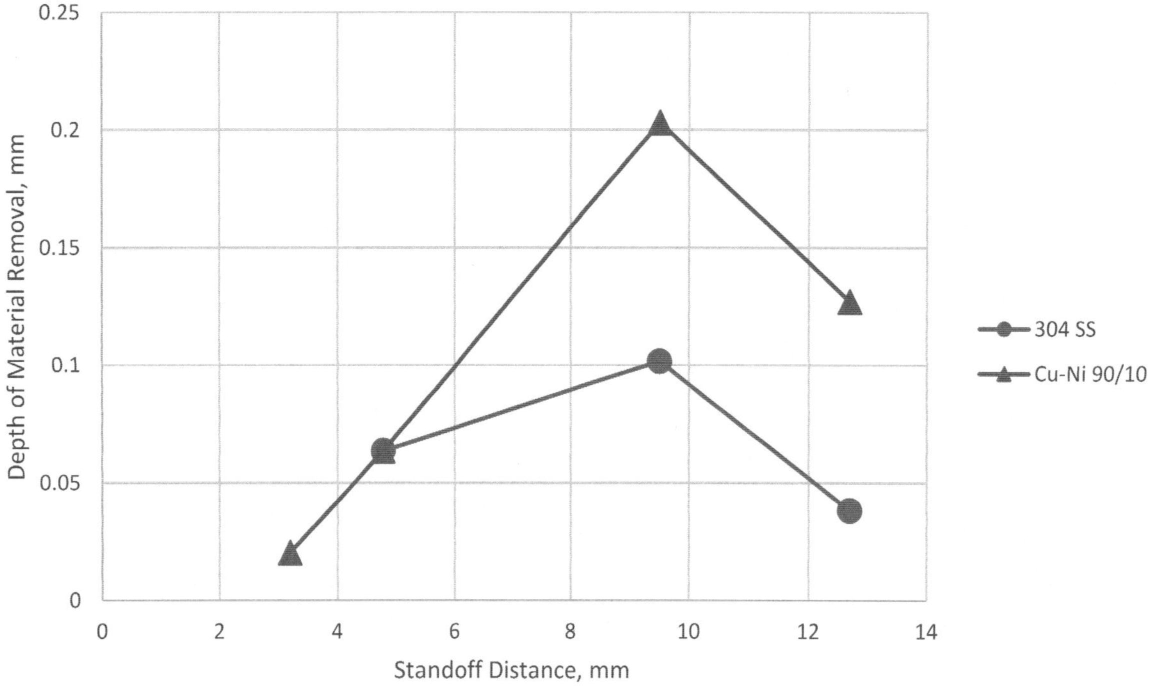

The Effect of Standoff Distance on Material Removal in 304 SS at 138 MPa (20,000 psi) and Copper-Nickel 90/10 at 69 MPa (10,000 psi) at 30 Second Exposure Times
Figure 12.

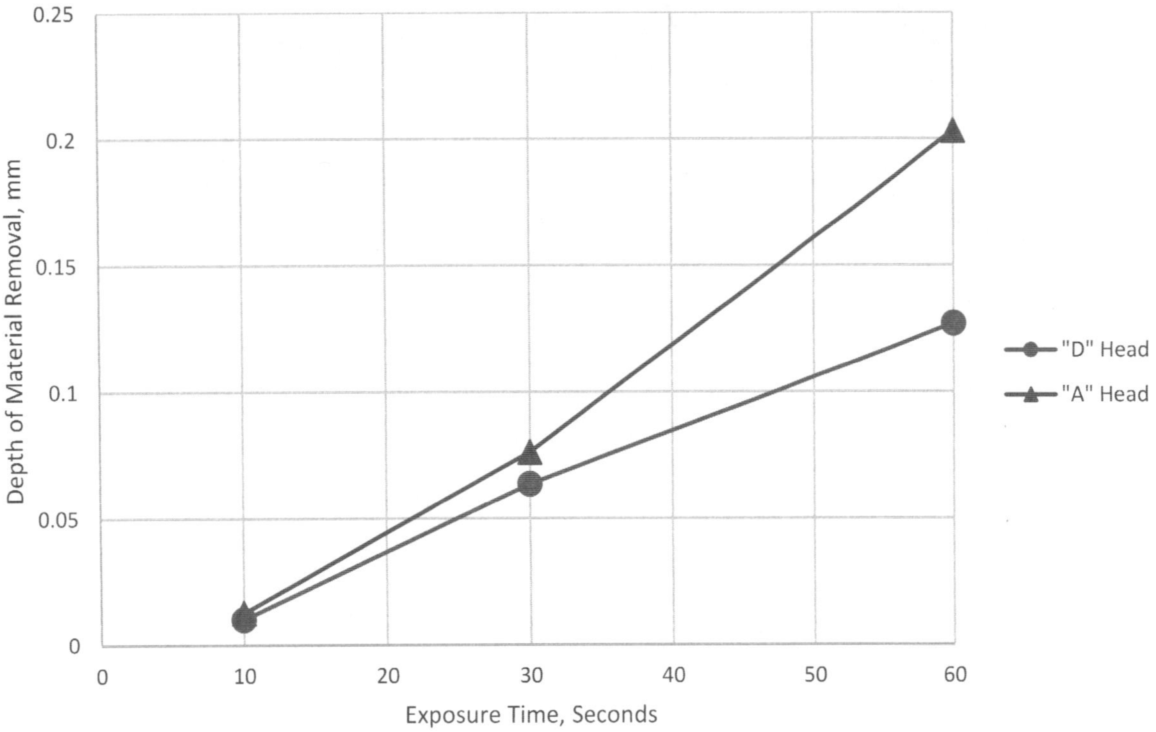

The Effect of Increasing Flow Rate in 1018 CS at 138 MPa (20,000 psi)
Figure 13.

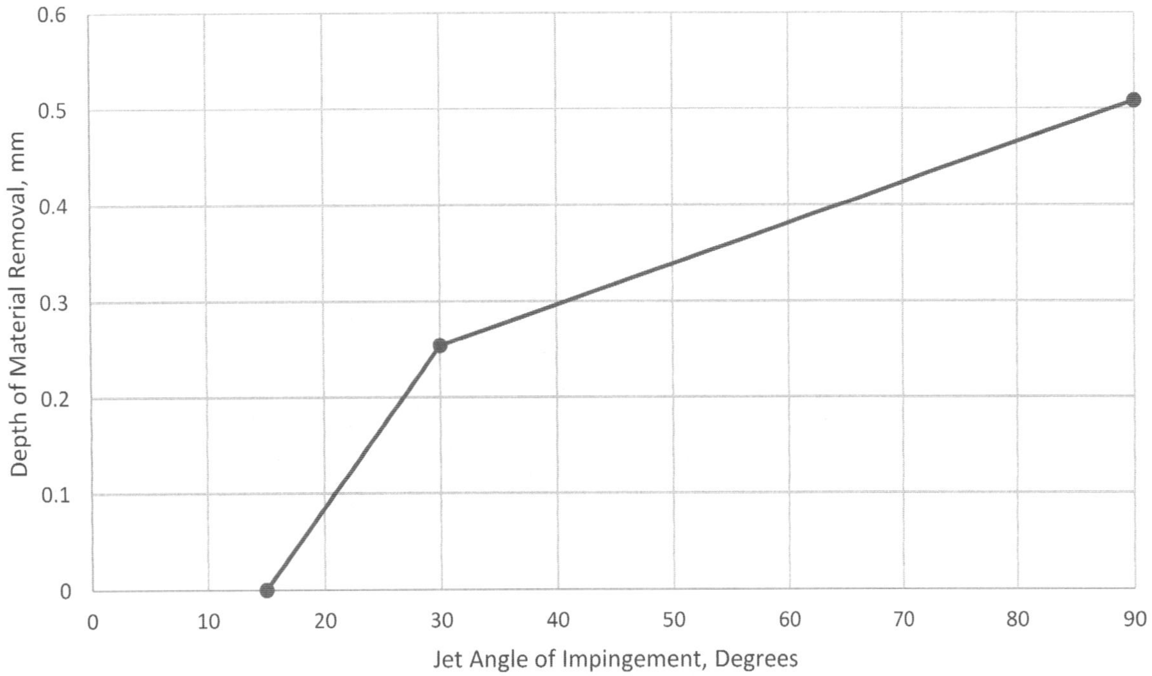

The Effect of Jet Angle of Impingement at 138 MPa (20,000 psi) in Copper 110 Plate
Figure 14.

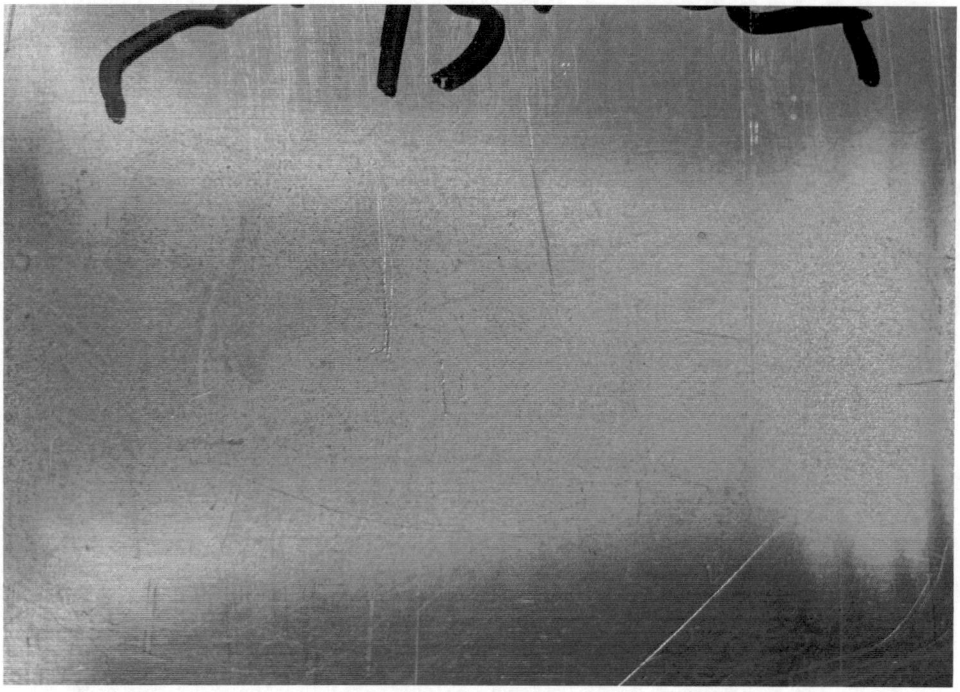

The Result of 100 Passes with Rotating Head at 103 MPa (15,000 psi) in Copper 110 Plate
Figure 15.

The Result of 50 Passes with Rotating Head at 228 MPa (33,000 psi) in 1018 CDF Plate
Figure 16.

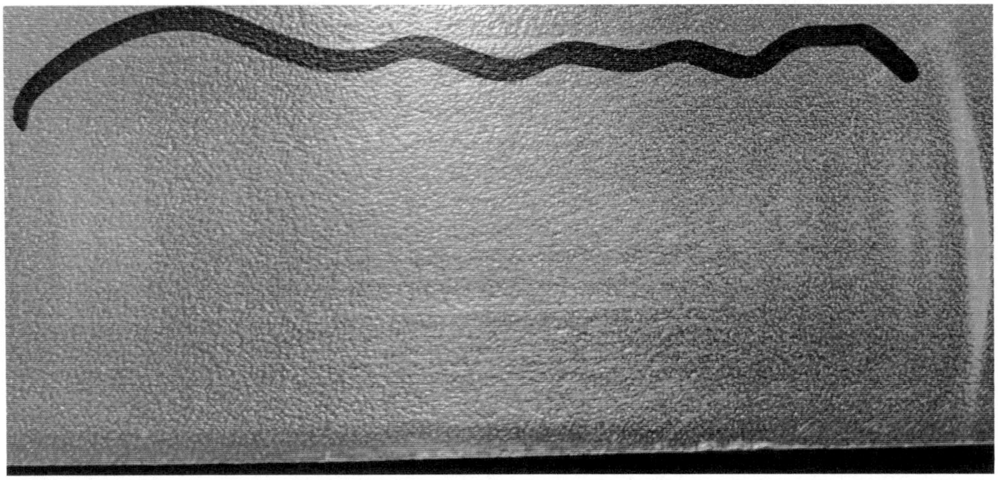

The Result of 50 Passes with Rotating Head at 228 MPa (33,000 psi) in Monel 400; Note Damage at End of Travel Range Where Pause of Direction Change Occurs
Figure 17.

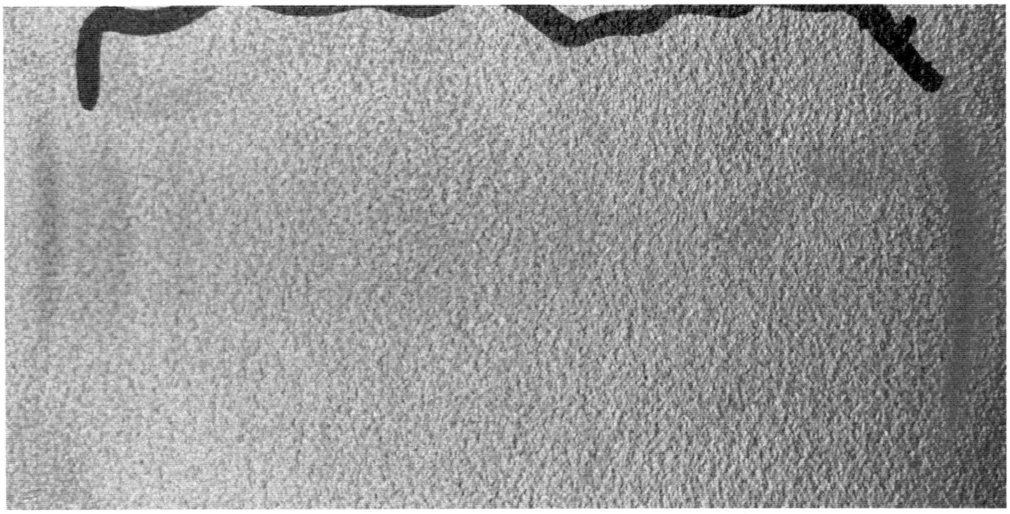

The Result of 50 Passes with Rotating Head at 228 MPa (33,000 psi) in Titanium
Figure 18.

2022 WJTA Conference and Expo
November 2-3 • New Orleans, Louisiana

Paper

DEVELOPMENTS IN PERSONAL SAFETY FOR INDUSTRIAL CLEANING

H. Borgt

The Dow Chemical Company

Terneuzen, The Netherlands

ABSTRACT

In the late 1970s, there has been a major shift in the way cleaning of industrial equipment was performed. Asset owners who have typically been performing the cleaning of their own equipment, based on cost reduction, started to outsource these activities to specialized industrial cleaning companies. Unfortunately, this transformation led to many severe safety incidents due to a lack of training/education, safe equipment, and best practices. To address this gap, safety associations were established in most industrialized countries. These associations developed best practices and guidelines for high-pressure water jetting and evolved to include other mechanical, chemical, and thermal methods, and related disciplines. Countries and regions that lacked these types of guidelines started to reach out to the established Industrial Cleaning Safety Associations.

To address these requests, the Global Industrial Cleaning Coalition (GICC) was established. The founding members, based on their experience, created seven basic principles of safe and effective cleaning to support the further development of best practices and safety associations in these underrepresented regions. In this paper, the history, and key drivers in the establishment of these safety associations, their output, and their benefits are summarized.

Key words: Personal safety, Industrial Cleaning

1 INTRODUCTION

"*Process equipment and piping in the chemical and petrochemical plants and petroleum refineries have to be cleaned periodically as part of normal maintenance operations to remove fouling that interferes with process flow, heat transfer, or other operations*"[1], different cleaning methods are being used to accomplish this. Currently, approximately 80% of all industrial cleaning activities are being performed by means of high-pressure water jetting, sometimes referred to as hydro blasting [15]. Due to increased expectations in quality, efficiency, and sustainability, new cleaning techniques have been developed (or existing techniques have been improved).

Performing industrial cleaning is not without risk. Operators performing the activity are exposed to the materials being removed during the cleaning process, and they are also at risk of safety incidents such as water jet cuts and abrasions. These incidents also occur in domestic use. "*An estimated 6,057 people in 2014 alone went to an emergency room with injuries related to pressure washer use, according to the Consumer Product Safety Commission. And 14 percent of those ER visits led to additional hospitalization.*" [2].

To address these incidents and risks, best practices and guidelines must be created that tackle the root causes of such incidents. To achieve that goal safety organizations have been founded around the globe. Dow participated in the founding of SIR, DIRV, EWJI, and GICC, and is an active member of WJTA.

2 HISTORY OF THE INDUSTRIAL CLEANING INDUSTRY

Looking back at the development of industrial cleaning in the petrochemical industry, it can be said that the first cleaning technologies were developed in the 1920s [1]. Some of them were mechanical but most of them chemical in nature. Due to side-effects (e.g., excessive corrosion, lack of effective cleaning techniques), more effort was put in optimizing current methods of cleaning and the development of other cleaning technologies. This approach led to the creation of service companies that provided contracting offerings, some of which were spin offs of asset owner companies, such as Dowell back in 1939.

In the 1950s, a new mechanical cleaning technology was introduced, water jetting, first as surface preparation method for cleaning surfaces and the removal of blistered paint. By increasing blasting pressures and flows, this method became suitable to remove deposits from tanks and other process equipment, which led to a widespread use for industrial cleaning purposes. In the 1960s, vacuum trucks were built to remove deposits from the bottom of the fire box in fired heaters, and sludge from the bottom of storage tanks, creating, together with water jetting, the foundation for the current industrial cleaning industry.

Since the 1970s, significant changes have occurred in equipment and piping cleaning. Environmental waste disposal and safety concerns have furthered the development of new cleaning methods and less hazardous cleaning chemicals.

"At this point in time, periodic cleaning of process equipment and piping has indeed become one of many necessary maintenance operations for basic chemical plants, gas plants, petroleum refineries, and petrochemical plants. Operating conditions have become more severe than previously. Process units have become more complex and are being operated over longer periods of time between scheduled turnarounds." [1]

This expansion of the industrial cleaning industry also led to an increase in incidents. The number and severity of these incidents, including fatalities, led to the formation of safety and technology associations in the 1980s. Refining (oil & gas) and petrochemicals where the main sectors that pushed for the formation of safety standards, requirements, and training.

2.1 Geographic Development

Comparing ethylene production and refinery capacities by country (as is shown in Figure 1) to the regions that have safety associations established or are in the process of establishing such organizations (Figure 2), a clear trend appears. This highlights the importance of these safety associations to these industries.

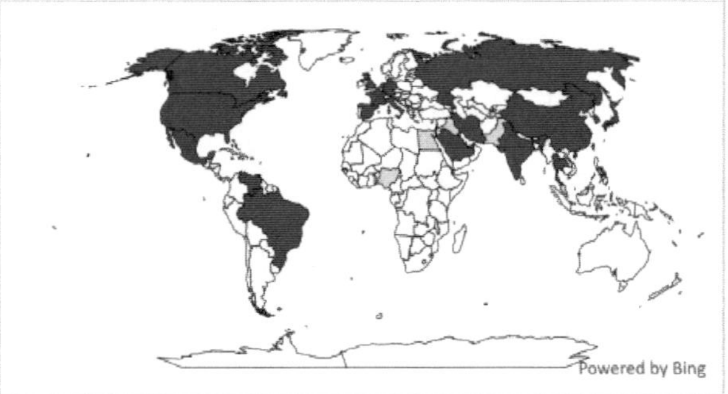

Figure 1. Countries with more than 1,000 ktpy Ethylene and more than 1 million BPD refinery capacity [8, 14]

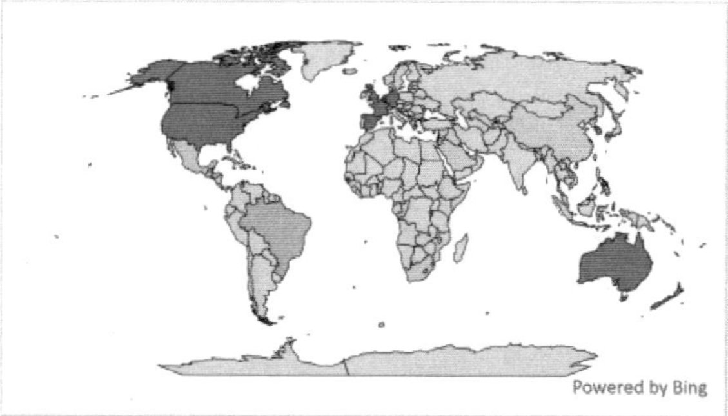

Figure 2. Countries with established safety associations (red), in process of establishing safety associations (green) and reached out for support (orange) [4, 5]

2.2 Safety associations

"*In a speech to the introductory session of the Congress, Dr. Jukka Takala, Chief of the ILO's Health and Safety program, pointed out that the workplace hecatomb of 1.1 million deaths exceeds the average annual deaths from road accidents (999,000), war (502,000), violence (563,000) and HIV/AIDS (312,000). .... In addition, he said that by conservative estimates workers suffer approximately 250 million occupational accidents and 160 million occupational diseases each year.*" [3]. The common ground of most of the safety organizations is that they have been founded as a direct result of serious incidents (fatalities, and severe incident trends). As an example, in the 80s, Stichting Industriele Reiniging (SIR) was founded after a streak of serious incidents that occurred in the Rotterdam, Netherlands area, resulting in multiple fatalities within one year. Incident review learned that lack of safe work procedures, guidelines, training, and equipment requirements where the main contributors. Reflecting on the last 15 years, SIR has seen a total of "only" four fatalities in this region. The Australasian High Pressure Water Jetting Association (AUSJET) was initially formed in the 90s, due to concerns within industry about the high rates of injuries sustained in the water jetting industry.

Globally more safety associations were formed with the same intent. A list with a short description of the leading safety and technology associations follows. The list is arranged chronologically, based on year of founding.

2.3 Water Jetting Association (WJA)

The Water Jetting Association (WJA) is the United Kingdom based trade association for businesses, agencies and individuals involved in the water jetting industry, with approximately 300 members. The WJA is the UK's largest provider of water jetting training. It has developed an internationally recognized Codes of Practice for water jetting which is continuously brought up-to-date and improved. The WJA was established in 1980 to represent members in the UK and is the oldest association in the world. It also welcomes international members and seeks to work with likeminded organizations across the world to further these goals. [11]

2.4 Water Jetting Technology Association (WJTA)

The WJTA is based in the United States and represents contractors, OEMs, asset owners, and training companies across North America. Its purpose is to promote and regulate safe working practices in industrial cleaning. WJTA collates and represents those interests in the US.

Mission & Vision:

The WJTA's mission to is advance safety and technology in industry, which is accomplished through establishing and promoting best practices; providing standardized operator training and certification; delivering education, news, and updates to association members; and organizing an annual trade show.

WJTA best practices for water jetting and industrial vacuuming are frequently reviewed by association committees to reflect advances in industry technology and safe work practices and to address identified safety hazards. Recent revisions to both manuals were published in 2021. In 2017 together with SIR and S3C founded the GICC.

2.5 Stichting Industriele Reiniging (SIR)

To promote and regulate safe working practices in industrial cleaning. SIR collates and represents those interests in the Netherlands and (since 2003) in Belgium. SIR is a textbook example of just how effective self-regulation of safety can be within a sector.

Mission & Vision:

Making industrial cleaning work as safe as possible: to achieve that objective, SIR is active in various areas. Setting guidelines is a key activity involving all parties from the sector. The unique makeup in terms of parties involved makes sure that the results obtained are balanced and backed by all those concerned. The guidelines that are defined apply throughout the sector in the Benelux countries. Every party that signs the SIR policy statement thereby undertakes to observe all the guidelines. The strength of the cooperation that is embedded in SIR is that current risks in industrial cleaning are acknowledged and appropriate guidelines for them accepted by all parties. Because the SIR guidelines are unambiguous, applying them reduces unsafe working situations to a minimum.

Industrial cleaning is an ever-changing discipline within which SIR always keeps its finger on the pulse. SIR has a motivating role in the development of safe equipment, and it is actively involved in developing safe working methods. It regularly adapts its guidelines to the latest state of the art. The network of steering committees and working groups containing representatives of various fields within the sector ensures that developments in the working domains are embodied in regulations, training, and examinations. The composition of these groups is always split proportionally between the business categories of SIR's participants so that the interests of all parties – which can sometimes be contradictory – can be properly considered for inclusion in the finalized guidelines. SIR is one of the founding members of GICC. Since it's founding SIR developed basic principles for safe working, GICCs seven basic principles are deviated from these principles.

2.6 AusJet

AUSJET is an Australian based safety association that promotes safe water jetting practices in Australasia.

Mission & Vision:

To develop a safe work environment through standards, training, and advocacy in the interest of our members throughout Australasia.

AUSJET and ADCVA endeavor to be the world leading association for the Water Jetting, Drain Cleaning, Vacuum Loading and Hydro Excavation Industries and lead in standards, safety, training, and industry advocacy. And it's core values are i) Cleaning-up Industry Safely, ii) 'Leadership – Zero Harm'

In 1999 two-joint Australian, New Zealand High Pressure Water Jetting standards were developed (AS/NZS 4233 1&2) and made available to industry. Due to the ever-changing nature of the industry, in 2022 AusJet has launched a project to facilitate a committee made up of member representatives to review the standard and implement changes to AS4233 Water Jetting Systems Operations (New Zealand no longer recognize this standard).

In 2013, high pressure water jetting's close association with drain cleaning, vacuum loading and hydro excavation services was recognized which also required the development of standards and training. This led to the decision to form the Australasian Drain Cleaning and Vacuum Association (ADCVA) and incorporate it with AUSJET.

Although the current training standards are recognized as a Nationally Accredited course and are mandatory to work on a high-pressure water jetting job, once the standard review is completed in 2022, AusJet will then review and update the training for water jetting, vacuum loading, drain cleaning and non-destructive digging to be formalized with a higher learning institute and aligning itself with other construction, manufacturing, and mining industry skills associations. [13]

2.7 Systeme de Certification, Competence et Conformite (S3C)

To promote and regulate safe working practices in industrial cleaning. S3C collates and represents those interests in France.

Mission & Vision:

Making industrial cleaning work as safe as possible: to achieve that objective, S3C focusses on the three C's: Certification, Competence and Conformity.

SIR, DIRV, and S3C have signed a mutual recognition agreement and are working closely together in the European Water Jet Institute to improve safety in the larger European area. [12]

2.8 Deutscher Industrie-Reinigungs Verband (DIRV)

DIRV is a German based safety association that promotes safe water jetting practices in Germany. [10]

Mission:

Work together with associated companies for adopting basic principles for safe industrial cleaning.

- To ensure a safe working place the latest developments of technology, working equipment and PPE need to be available to protect employees from risks. These developments need to be implemented by companies working together.
- To ensure qualified employees professional trainings and certification need to be conducted, which empowers the people to raise their working skills to a high level of quality and what is necessary to decide on how to deliver the best results for the task as required.

Vision:

The vision of the DIRV is to implement the safest possible industrial cleaning standards in Germany and to ensure that people doing their work in safe environment. This will be achieved by: Professional trained and certified labor, Setting standards for occupational safety and PPE, Setting standards for highly developed equipment, Working together with all different cleaning companies for achieving the best available cleaning results, Setting standards for implementing international guidelines and requirements according to German legal requirements, Working in communities for the exchange of lessons learned

2.9 European Water Jetting Institute (EWJI)

Both the Global Industrial Cleaning Coalition (GICC) and the European Water Jetting Institute (EWJI) are not country-based but work as a forum for the other organizations to interact, streamline, and harmonize their requirements and best practices.

EWJI, as a non-profit organization, has the following objectives:

To be involved and have an impact on health and safety standards, best practices, training, and legislation related with water jetting. To promote European regulations and reference documents on water jetting techniques, to improve processes and ensure quality. To create a network for open communication and exchange among water jetting professionals, companies, associations, and related stakeholders. To exchange information on techniques, applications and working methods among the industry and with other related industries. To foster innovation, through research and cooperation between the industry and universities, and to facilitate the access to funding available for these purposes. To support the existing associations and foster the creation of new organizations in Europe, sharing the knowledge and best practices related with water jetting. To contact and cooperate with similar organizations in other parts of the world.

2.10 Global Industrial Cleaning Coalition (GICC)

In 2017 SIR, WJTA and S3C collaborated in the formation of the GICC. This forum aims to assist countries and regions where safety associations lack to provide guidance in forming such organizations. At the foundation of this forum, the coalition published the seven basic principles of industrial cleaning which embraces the principles of all three organizations. Besides providing support for the developing areas this forum provides a platform where the different safety associations can exchange experience and information, to improve safety in the global industrial cleaning industry, without impacting the sovereignty of the specific organizations.

Looking at the scope of the safety organizations it shows that high pressure water jetting activities and vacuum loading are best served. It should be noted that all the safety associations mentioned above are non-profit. Additionally, active associations are: ALTAP (Spain), Korean Water Jetting Association (South Korea) are also active. The Industrial Cleaning Association China (China) is currently being established.

Table 1. Scope and focus of safety and technology associations [4, 5, 7, 10, 11, 12, 13]

Safety Association		Scope					Focus				
Name	Founded	Water Jetting	Vacuum	Chemical	Other	Respiratory Air	Petro- Chemical	Municipalities	Construction/Demolition	Mining/Minerals	Other Industry
WJA	1980	x	x				x	x	x		x
WJTA	1983	x	x	x			x	x	x	x	x
SIR	1987	x	x	x		x	x	x			x
AUSJET	1991	x	x				x	x	x	x	x
S3C	2008	x	x				x	x			x
EWJI	2015	x	x				x	x	x		x
GICC	2017	x	x	x			x				x
DIRV	2019	x	x	x	x		x	x	x		x

3 PERSONAL SAFETY INCIDENTS

Personal safety is paramount in industrial cleaning. Unfortunate, industrial cleaning activities have a high injury potential: people can be struck by jets or abrasives leading to damaged skin, tissue, broken bones, contamination with fouling products and other more generic risks. Currently, most of the focus in industry with respect to cleaning safety is related to high pressure water jetting. However, other cleaning disciplines have a similar or maybe even greater potential towards number of incidents and their severity.

3.1 Incident data review

Data resources for Industrial Cleaning incidents are scarce, for this paper the following sources where used:

- SIR Incident database 2010-2021
- AusJet Waterjet incidents 1998-2012

The benefit of the SIR data is that it creates a better and broader understanding of the incidents that have occurred in the larger industrial cleaning space, while the data from AusJet solely focus on high pressure water jetting.

If we look at the number of incidents per discipline it shows that vacuum loading has a greater impact than high pressure cleaning followed by respiratory air protection and chemical cleaning.[1]

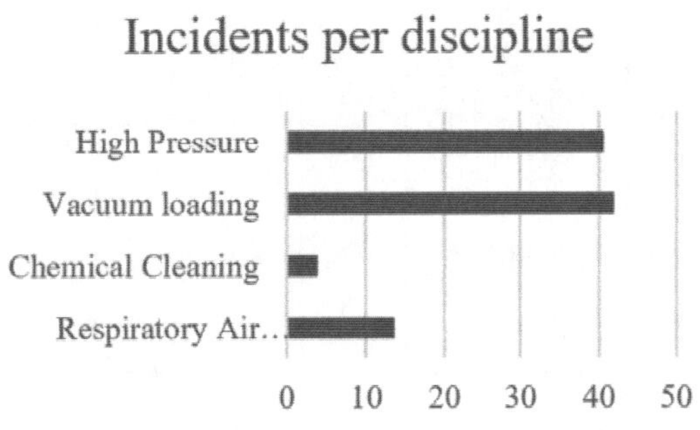

Figure 3. Incidents per discipline [9]

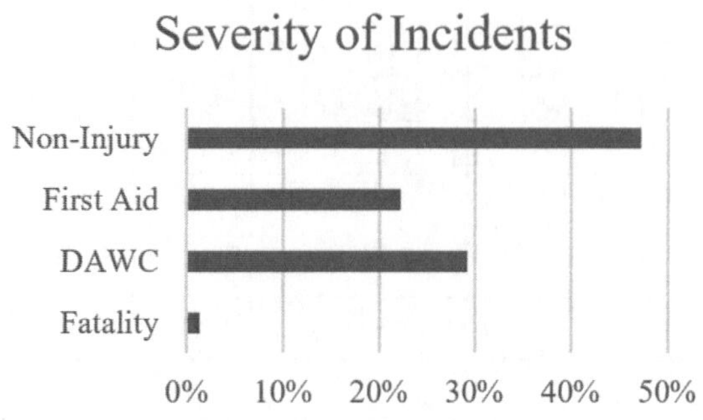

Figure 4. Severity of incidents [9]

A deeper dive in the same incident data show the following: fatalities cover about 1.5% of all incidents, day away of work cases 29%, first aid cases 22% and non-injury (near misses and damaged cleaning equipment) for 47%.

The fact that the Day Away of Work Cases (DAWC's) are more common than the first aid cases is interesting to notice, since the safety pyramid would normally say otherwise. This shows the severity potential of getting injured during high pressure cleaning activities.

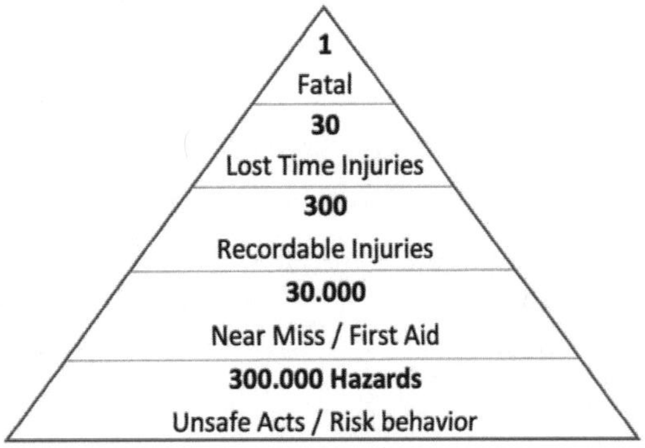

Figure 5. Illustration of Heinirch's theory "the Safety pyramid" [16]

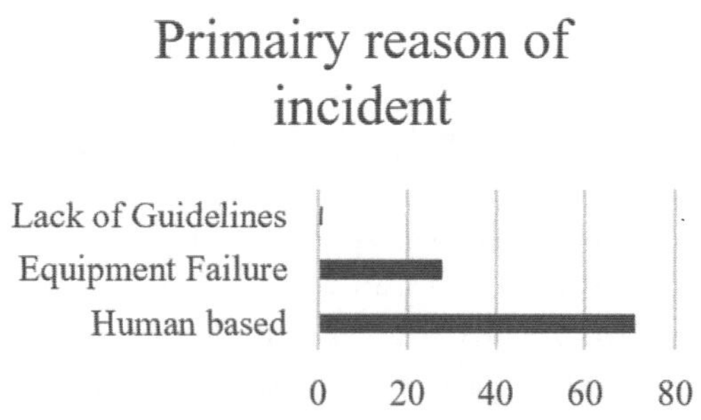

Figure 6. Primary reasons of incidents [9]

Continuing the deep dive into the data, the primary reasons of incidents have been identified i) human behavior, accounting for ca. 70% of the incidents, ii) equipment failure, accounting for ca 30% and iii) the lack of best practices, regulation or guidelines which accounts for about 1%.[2] The following sections provide more insights for each of these factors.

[1] The magnitude of the vacuum loading and high-pressure activities is considerably larger than the chemical cleaning activities.

[2] Take into account that SIR created and published its guidelines and requirement well ahead 2010.

3.1.1 The human based factor

The human based factor consists out of two components i) experience/awareness and ii) training. The AusJet data provides an insight in the experience part of this factor.

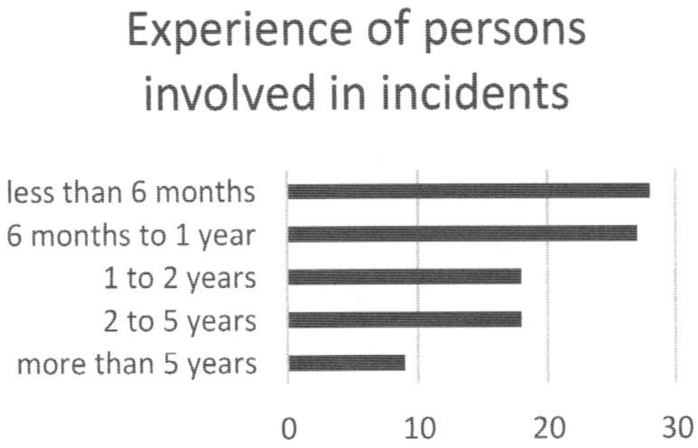

Figure 7. Prior experience of operators involved in incidents [13]

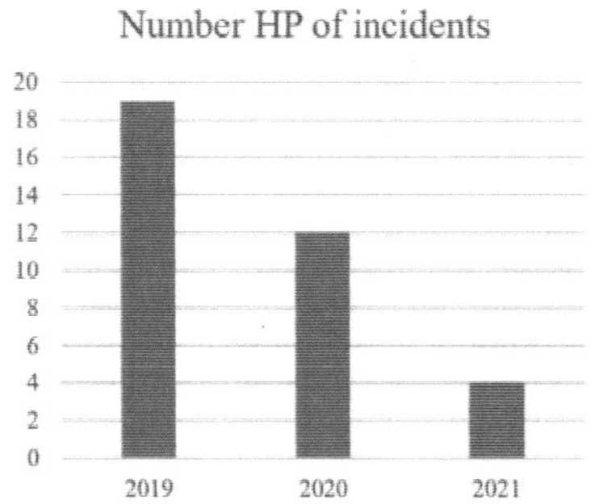

Figure 8. High pressure water jetting incidents Netherlands and Belgium [9]

The number of incidents where the experience of the operator is less than a year drastically drops and drops even further for operators with more than 5 year experience which clearly shows the importance for training and experience. Incidents were operators with more than 5 years of experience are involved seem of a higher severity rate and seem to be related to complacency.

3.1.2 The equipment-based factor

Equipment failures consist mainly of connections and hose failing. In the vacuum loading space mainly design issues. Most of these design issues have been addressed in updated guidelines and requirements.

3.1.3 Lack of best practices

Data for regions where guidelines and best practices are lacking, shows a larger incident rate. This is mainly the reason of reach-out from regions like China, South Africa, Saudi Arabia, Mexico, and Brazil. Most of these initiatives are driven from asset owners.

To address these issues, the GICC, made up of the standing safety/technology associations, worked together to create basic principles as a blueprint to adhere to. These seven basic principles are further discussed in this paper.

The effect of best practices can be shown with the following example. Through an intensive collaboration between asset owners, industrial cleaning companies and manufacturers/suppliers. Since 2020 Non-manual high-pressure cleaning is the new standard which is being achieved by automated and mechanical jetting by SIR in the Netherlands and Belgium, which led to a further decrease in high pressure water jetting incidents.

The incident rate in comparison to the number of activities and performing hours was already very low, compared with the global incident rates for high-pressure cleaning.

3.2 Expansion to other disciplines

Almost all associations originally started with high pressure cleaning and then expanded based on a demand from industry. In the last few years, more focus has been given to subjects such as sustainability, efficiency, and waste, leading to the development and re-invention of other cleaning technologies.

These other technologies require their own set of guidelines and best practices.

4 Seven basic principles

Frameworks are important tools that help to improve safety and avoid incidents described above, the Global Industrial Cleaning Coalition introduced the seven basic principles for industrial cleaning safety associations.

These principles provide a blueprint for new to create safety standards.

The seven basic principles are:

1. Joint Responsibility
2. Three Areas of Focus
3. Education & Training
4. Knowledge & Learning
5. Assurance of Guidelines
6. Working with Safe Equipment
7. Working According to Safe Methods

These 7 principles are applicable on all different methods of cleaning, at this moment the following 3 areas have been expanded:

- HP Water jetting
- Industrial Vacuuming
- Chemical Cleaning

Principle 1 to 5 are equal for all different cleaning methods and principle 6 and 7 are tailored to the specific technology. [4]

The guidelines and best practices of all forum members have been reviewed by a technical committee and where required an implementation plan has put in place to improve the guidelines and best practices.

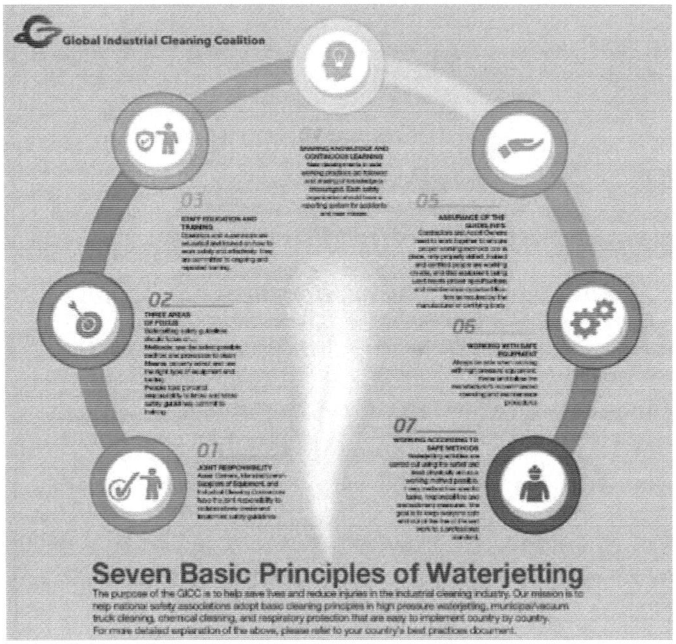

Figure 9. Schematic overview of the Seven Basic Principles of Water Jetting [4]

4.1 Joint Responsibility

Asset Owners, Manufacturers/-Suppliers of Equipment, and Industrial Cleaning Contractors have a joint responsibility to collaboratively create and implement safety guidelines

It is obvious that the occurrence of incidents and accidents is a serious matter, and that effective control of the risks requires cooperation from all parties. That makes it important that the specific responsibilities are known.

The Asset Owner's policy must focus on making sure that cleaning activities can be carried out under conditions in which the employee is not directly exposed to the associated risks. This means that the client creates conditions where reasonably possible in which the cleaning can be carried out in a fully automatic, mechanized way or in a mechanized way that is controlled from outside the cleaning area or inside if necessary. The client must consult the cleaning company about this.

Industrial cleaning companies are responsible for carrying out cleaning activities correctly. The employer is also responsible for the well-being and safety of employees who carry out cleaning activities. The legal regulations about working conditions must be observed.

The task of the manufacturer/supplier is to inform and advise the cleaning company and other parties involved in the best possible way about the proper selection, use and maintenance of a cleaning system and/or components. The manufacturer/supplier provides clear user instructions. At the customer's request, the manufacturer or supplier trains or instructs the users and provides a personalized document as evidence showing that the person in question has been trained or instructed in using the cleaning system and/or cleaning components. Any defects in the cleaning equipment are immediately reported to the customers of the equipment.

4.2 Three Areas of Focus

Industrial Cleaning safety guidelines should focus on…

Methods: use the safest possible method and processes to clean.

Means: properly select and use the right type of equipment and tooling.

People take personal responsibility to know and follow safety guidelines; commit to training.

Methods: safe working methods that apply to the client industries, the cleaning companies doing the work, the inspection bodies and other companies that have included the safety guidelines in their safety policy and business operations.

Means: equipment that must comply with the requirements when first used and that must continue to comply during its lifespan. This can be determined by means of regular inspection and testing.

People: staff who remain professional thanks to education, training examination based on the guidelines from the safety association.

4.3 Education & Training

Operators and supervisors are educated and trained on how to work safely and effectively. They are committed to ongoing and repeated training.

As showed in the incident data review, most of the incidents are human related. By providing repeated safety and awareness training and instruction, this risk can be addressed.

4.4 Knowledge & Learning

New developments in safe working practices are followed and sharing of knowledge is encouraged. Each safety organization should have a reporting system for accidents and near misses.

It's important that the complete industry learns from incidents and issues. Tracking incidents, providing lessons learned and adjusting best practices and guidelines based on that experience will improve the complete industry.

4.5 Assurance of Guidelines

Contractors and Asset Owners need to work together to ensure proper working methods are in place, only properly skilled, trained, and certified people are working on-site, and that equipment being used meets proper specifications and maintenance cycle/certifications as required by the manufacturer or certifying body.

People: Training certification based on the guidelines from the safety association.

Means: Test certificates are issued on approval that are valid for a certain period.

Working with Safe Equipment

Always be safe when working with cleaning equipment. Know and follow the manufacturer's recommended operating and maintenance procedures.

Based on experiences and incidents in the past guidelines have been created for equipment, for example EN-1829 documents. These documents deal with all significant hazards, hazardous situations and events arising during assembly, erection, operation and servicing relevant to high pressure water jet machines (EN-1829-1) and hoses (EN 1829-2). Now a harmonization and streamlining project is ongoing with IEC SC61J to create a global/international standard.

4.6 Working According to Safe Methods

Cleaning activities are carried out using the safest and least physically arduous working method possible. Every method has specific tasks, responsibilities, and precautionary measures. The goal is to keep everyone safe and out of the line of fire and work to a professional standard.

Following the hierarchy of control will help with the reduction and severity of incidents. Eliminating the hazard and risk is the highest level of control, followed by reducing the risk through engineered solutions. If these don't provide sufficient mitigation of the risks, administrative controls follow. Reducing the risk using protective personal equipment (PPE) is the lowest level of control.

5 CONCLUSION

Most of the traditional safety/technology associations have been formed between 20 and 40 years ago, as a response to the large number of severe incidents experienced in the refining and petrochemical industries. The regions where such organizations are operating and have been embraced by governments and local industry see a steep decline in these severe incidents.

When following GICC's seven basic principles a clear improvement of safety is being experienced.

6 References

[1] Gutzeit, J. (1997) *Cleaning of Process Equipment and Piping*, St. Louis, MO, Materials Technology Institute.

[2] Perratore, E. (08 March 2016) Safety alert under pressure, Consumer Reports. http://www.consumerreports.org/pressure-washers/safety-alert-under-pressure/ [Accessed 31 January 2022].

[3] International Labour Organization. (1999) *ILO Estimates Over 1 Million Work-Related Fatalities Each Year*. https://www.ilo.org/global/about-the-ilo/newsroom/news/WCMS_007969/lang--en/index.htm [Accessed 31 January 2022]

[4] Global Industrial Cleaning Coalition (2021) *Cleaning Principles.* https://www.globalindustrialcleaningcoalition.org/cleaning-principles/hp-waterjetting [Accessed 3 February 2022]

[5] European Water Jetting Institute (2021) *Institute.* https://www.ewji.org/institute/presentation [Accessed 2 February 2022]

[6] Momber, A. (2003), Hydroblasting & Coating of Steel Structures, Oxford, Elsevier.

[7] van Teijlingen, Martin, Mission and vision of SIR, personal communication, March 2022

[8] Oil & Gas Journal (2010), *FOCUS: Global Ethylene Production.* http://www.ogj.com/home/article/17209147/OGJ-FOCUSGLOBAL-ETHYLENE-PRODUCTION [Accessed 4 February 2022]

[9] Stichting Industriele Reiniging, *Incident database 2010-2021.,* Januari 2022.

[10] Felske, Andreas, Mission and vision of DIRV, personal communication, March 2022

[11] Wilson, Steve, Mission and vision of WJA, personal communication, March 2022

[12] Harwood, Stuart, Mission and vision of S3C, personal communication, March 2022

[13] McKay, Lance. Mission and vision of AusJet, personal communication, March 2022

[14] Wikipedia (2010), List of oil refineries. http://en.wikipedia.org/wiki/List_of_oil_refineries [Accessed 4 February 2022]

[15] Industry Report 2021, European Water Jetting Institute, November 2021

[16] Resource Management Inc., Workplace Safety Tip, RMI Client Newsletter: January 2010 (rminc.com) [Accessed 6 February 2022]

2022 WJTA Conference and Expo
November 2-3 • New Orleans, Louisiana

Paper

UTILIZING WATERJET TECHNOLOGY TO MITIGATE SUBSTRATE DAMAGE WITHIN UNDERGROUND STRUCTURAL LINERS DURING REHABILITATION WORK

J. Bourgeois, H. Miller, E. Charrier
Colorado School of Mines
Golden, CO

ABSTRACT

Waterjet excavation has the potential to eliminate many of the technical and operating challenges associated with conventional shotcrete removal and repair. Evidence derived through empiric laboratory testing indicates that waterjets are capable of selectively removing damaged areas of support liners without structurally compromising the substrate and adjacent intact sections of the liner. This research illustrates the contrast between the excavation process associated with both conventional mechanical impact hammers and waterjet excavation methods during empiric testing. An analysis on the fracture mechanisms and operating parameters of each method was completed. Within this analysis, instrumented shotcrete panels were physically tested to quantify vibration during excavation and examined through visual and analytic processes to determine substrate damage and delamination. After testing was completed, this data strongly indicated that waterjet cutting causes less collateral damage to the surrounding intact liner and substrate when compared to that of conventional impact hammers. This research was conducted to provide a scientific basis for additional applied research in the rapid excavation and repair of shotcrete support systems. The intent is to develop a viable technology that will reduce the collateral damage caused to surrounding liners during excavation, improve the adhesion between shotcrete and the substrate for longer lasting ground support, and improve overall safety for workers in underground environments.

Organized and Sponsored by WJTA®

1. INTRODUCTION

The repair of concrete and shotcrete liners that have been structurally compromised or damaged is a common maintenance and rehabilitation activity in tunnels and other types of underground workings. In many cases, it is prudent to limit the repair to the isolated removal of the structural liner around the damaged area rather than the complete excavation of large sections of the support system. In these applications, traditional methods of liner removal include the use of hydraulic or pneumatic hammers that break and excavate the liner material through repetitive percussive impacts. In operating environments where working heights are less than twenty meters, these hammers are usually mounted to articulated booms attached to mobile rubber-tired or track equipment, such as a mechanical scaler. In applications that extend beyond the reach of conventional scalers, the height of the operating envelope usually necessitates the use of man-lifts, where workers manually remove the damaged areas from these elevated platforms using hand-held equipment.

There are significant drawbacks to the use of these traditional excavation methods. It is a long-held belief that the percussive impacts and vibration generated by hydraulic/pneumatic hammers while removing the compromised area of the liner also causes unintentional damage to the surrounding intact shotcrete and/or concrete because of the propagation of fractures and the delamination of the liner from the rock substrate, as well as any contained rebar or wire mesh/screen backing. In addition, there is a host of potential safety and health hazards commonly associated with the use of these traditional technologies. This is particularly true for work performed from elevated platforms, including the worker's close proximity to unstable roof/back, the potential of falling from these platforms, the man-lift tipping or overturning, and the limited ability for rapid egress.

In addition to these drawbacks, it is important to recognize that there is significant potential for growth in underground rehabilitation work due to current operating trends in both the mining and tunneling industries. Given the cumulative impact of increasing depth, age, in-situ stresses, and complex operating environments, higher demands on ground support, specifically through the use of shotcrete, are necessary to reduce instances of ground falls and support failure. The resulting high demand of ground support in the rehabilitation of existing underground workings will encompass new challenges that require the development of safe and efficient shotcrete rehabilitation practices and technologies that can be remotely operated in a wide range of work environments [1].

Building upon the success of previous Colorado School of Mines (CSM) research activities in underground rock scaling and scarification [2], evidence from this study indicates that waterjets are capable of selectively removing damaged areas of support liners without structurally compromising or adversely impacting intact material around the area being repaired. This hypothesis is partially supported using hydrodemolition on deteriorated concrete structures [3] with a specific focus on its application on bridge decks [4] [5]. However, due to the differences in excavation layout, operating characteristics, and environmental conditions between bridge deck removal and mine drift/tunnel liner rehabilitation, the intent of this research is to facilitate a baseline understanding of the excavation processes associated with liner repair in hopes of developing a future prototype system applicable for field testing.

2. MATERIALS AND METHODS

This analysis is founded on empiric cutting tests performed on engineered and instrumented shotcrete panels designed to facilitate qualitative comparison of fracture propagation, substrate delamination, and collateral damage associated with waterjet and impact hammer excavation. Under the testing methodology, seven individual composite panels comprised of a steel reinforced concrete substrate with a 10-centimeter (4-inch) shotcrete layer were developed at the Earth Mechanics Institute (EMI) at the CSM in Golden, CO.

As part of the research testing plan, six composite panels underwent physical testing; where three of these panels were cut by a twin-orifice, rotating waterjet and three panels were mechanically excavated by a pneumatic percussive impact hammer. One panel was held in reserve for precautionary reasons. For all testing scenarios, the panels were excavated until the tool cut through the entirety of the shotcrete layer, reaching the interface between shotcrete and concrete. Each panel was instrumented with accelerometers attached to the rebar that reinforced the concrete to collect vibration data during testing. After testing, panels were visually examined to identify instances of surface fracture networks, delamination, and split aggregate from the excavation tool. Non-destructive structural analysis of the panels was performed before and after empiric testing using ground penetrating radar (GPR). The data collected through GPR will not be discussed in the context of this paper, however it can be referenced in Josef Bourgeois' dissertation "Characterizing the excavation Process Associated with Removal of Shotcrete Reinforced Underground Structural Liners Using Waterjet Technology" [6].

2.1 Engineered Test Panel

The test panels measure 76 centimeters (30 inches) by 61 centimeters (24 inches) by 25 centimeters (10 inches) thick, weighing approximately 226.8 kilograms (500 pounds) per panel. The concrete substrate is 15 centimeters (6 inches) thick and reinforced with #3 rebar (0.95 centimeters or 3/8 inches). The rebar is positioned in the middle of the concrete layer in a cross pattern on 20-centimeter (8-inch) spacing. Three segments of rebar are oriented along the short-axis, while two rebar segments run along the long-axis in each sample. The concrete/shotcrete and steel reinforcing materials are housed by a wood panel form. Holes were drilled in the sides of the wooden panel form to hold the rebar at the appropriate location within the panel, 7.6 centimeters (3 inches) up from the plywood bottom. The rebar was placed in the holes and protruded out of each panel siding to mount the accelerometers for vibration measurements. The concrete is type I/II 31 MPa (4,500 PSI (min)) commercial ready mix with 2-centimeter (¾-inch) aggregate. The batch was mixed and delivered by a local ready-mix company to the EMI and poured in the seven wood panel forms. A standard electric vibrator was used on the concrete after it was poured to eliminate the potential for air voids in the batch within each panel.

After the concrete cured for four weeks, the shotcrete layer was poured. The shotcrete layer is 10 centimeters (4 inches) thick with a 4-gauge welded wire mesh positioned 5 centimeters (2 inches) above the substrate. To facilitate the use of GPR in damage assessment and provide uniform shotcrete layering on the test panels, Quikrete sand/topping mix was used as a shotcrete simulant. This was necessary because the size of the test panels does not allow for consistent placement of shotcrete from a pneumatic sprayer. The steel (rebar and wire mesh) placement and density as well as the materials used in the substrate and shotcrete simulant were selected to replicate design parameters consistent with underground support liners. Panel dimensions (in inches) are shown in Figure 1 and panel properties are outlined in Table 1.

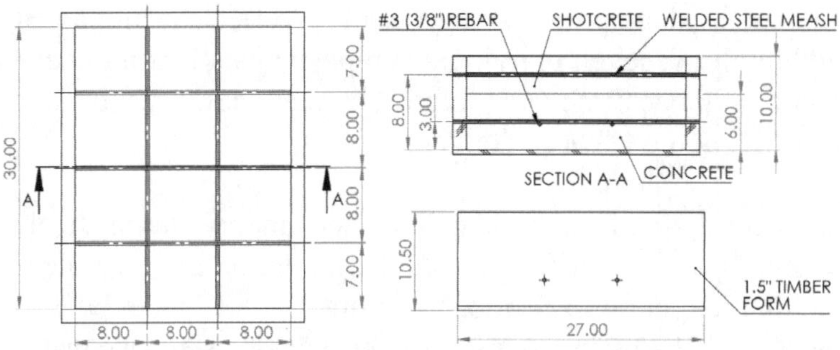

Figure 1. Composite panel dimensions (shown in inches)

Table 1.Composite panel properties

Panel Component	*Material Volume, cm³ (in³)*	*% of Panel*
Type I/II 4,500 PSI Concrete	70,553 (4,305.4)	59.8
Quikrete Sand/Topping Mix	47,111 (2,874.9)	39.9
#3 (3/8") Rebar	239 (14.6)	0.2
Welded Steel Mesh	84 (5.1)	0.1
Sum	**117,987 (7,200)**	**100**

*Welded steel mesh nominal diameter of 4 gauge – 4.7 mm (0.187")

2.2 Mechanical Excavation

Mechanical excavation was conducted at the EMI using an electric rotary hammer. The sample was turned on end, so the exposed concrete face was vertical in order to better simulate repairs to the rib. A Bauer 40-millimeter (1-9/16-inch) rotary hammer (1643E-B) was used. It was rated at 3,780 impacts per minute with a force of up to 6.6 foot-pounds per impact. Excavation was conducted using a 2.54-centimeter (1-inch) concrete spade chisel. Bit rotation was disabled to allow for better operator control of the chisel location and orientation and to simulate common repair practice. A rectangular cavity was cut in the center of each panel and excavated to the top of the concrete layer. The mechanical excavation process is shown in Figure 2.

Figure 2. Process of mechanical excavation using impact hammer

2.3 Waterjet Excavation

The panels that underwent waterjet testing were placed on a manually operated hydraulic lift to be positioned similar to that of the panels tested by the impact hammer. This provided vertical movement of the sample and was utilized to elevate the panel relative to the jet assembly. Horizontal movement during sample cutting was accomplished by a pneumatic actuated linear slide. The pressurized fluid flow for waterjet testing was generated by a KOBE high-pressure triplex, positive displacement pump. This pump was a Size-4 model with 180 nominal HP and 15.9-millimeter (5/8-inch) diameter plungers. It possessed a designed maximum operating pressure of 207 MPa (30,000 PSI) as well as a theoretical fixed displacement of 28.1 lpm (7.4 gpm). The KOBE was powered by a 150 HP electric AC motor. During testing, the system operating pressure was 189.6 MPa (27,500 PSI), which was measured at the pump discharge manifold. Due to the occurrence of slippage internally within the pump at pressure, the actual nozzle flow rate during testing was approximately 21.3 lpm (5.6 gpm) rate. During operation of the waterjet system for testing, a low-pressure priming pump was used to supply the Kobe pump at a flow rate of 22.8 lpm (6 gpm) at 0.45 MPa (65 PSI). The system also included a 950-liter (250-gallon) water storage tank and inlet line filters. The waterjet system diagram is shown in Figure 3.

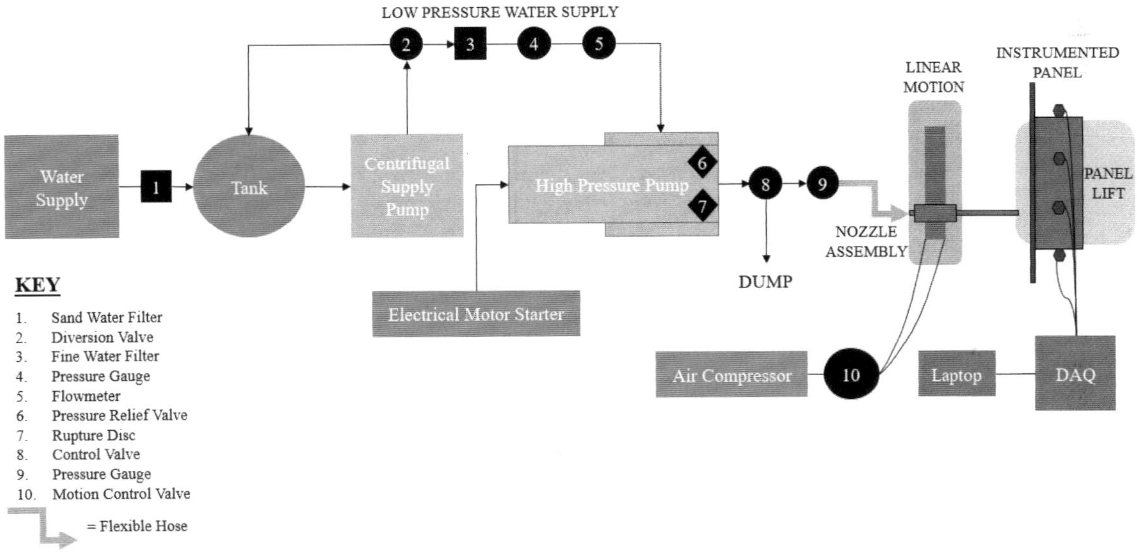

Figure 3. Waterjet instrumentation and cutting setup

A twin orifice rotary nozzle assembly was utilized with two commercially purchased 0.74 millimeter (0.029 inch) diameter stainless steel leach walker type nozzles with vane flow straighteners. These nozzles were chosen based on their pressure and flow ratings of 151.7 MPa (22,000 PSI) and 12.9 lpm (3.4 gpm) respectively. Overall, the size and design of the nozzles used for this research was chosen based on empiric experience and the need to maintain jet stagnation pressures above the potential threshold pressures of the target material. The nozzle assembly was connected to the pump accumulator via a 3.7-meter (12-foot) long, 5.1-millimeter (0.2 inch) ID, 275.8 MPa (40,000 PSI) high pressure flexible hose. The system includes a bypass valve at the pump which was designed to regulate flow and pressure. For safety considerations, a pressure relief valve and rupture disc assembly were plumbed into the system. System pressure was measured using a gauge located at the pump accumulator. The pump required an organic lubricant in the form of a soluble gear oil mixture, which was mixed into the water supply (2% by volume). It is assumed that the addition of the oil into the system possessed negligible effects on the thermal and dynamic properties of the fluid. A summary of operating parameters used during continuous waterjet impingement on the test panels is outlined in Table 2. The cutting system used during waterjet testing is shown in Figure 4.

Table 2. Continuous waterjet testing parameters

Jet Parameter	Value	Units
Fluid Pressure at Pump	27,500	PSI
Fluid Pressure at Nozzle	25,966	PSI
System Pressure Loss	1,534	PSI
Pump Flow	5.6	GPM
Nozzle Diameter	0.029	in
Nozzle Exit Velocity	1,768	ft/s
Standoff Distance	3.5 – 7.5	in
Traverse Rate	1.0 – 6.0	in/s
Impingement Angle	10	deg
Rotation of Nozzle Assembly	300	RPM

Figure 4. Waterjet excavation – Laboratory set-up (upper left) start of slot in panel (upper right) kerf in panel (bottom left) close-up of kerf in panel (bottom right)

2.4 Visual Inspection

Visual inspection was performed after excavation was completed on each panel. This process was accomplished in sections to analyze the entirety of the excavation profile including the left, upper, right, and bottom walls, as well as the face where exposed aggregate was located. Before inspection took place, each panel's excavation profile was rinsed thoroughly with water to provide a clean, wet surface which aided in fracture/failure visibility to the naked eye. When each section was examined, areas of fracture/failure were noted. These areas either had a single fracture or a network of failure which were manually measured in millimeters using measuring tape. All these areas were identified, measured, and summarized for each excavation section. During inspection of a particular section, if damage to wire mesh was present, it was noted for final comparison between each testing method. Delamination was also measured in millimeters if it was present in the analyzed division of the excavation profile. Finally, for the face section, split aggregate was counted and divided by the total exposed aggregate to give a percentage of split aggregate at the face due to testing.

2.5 Vibration Data Acquisition

The data acquisition system (DAQ) recorded vibration readings from seven accelerometers. The accelerometers were mounted on the exposed ends of the rebar in the test panels. A USB DAQ and LabVIEW were used to process and save the collected data to a laptop. The accelerometers used were a prosumer product from Sparkfun, the ADXL337, which is a triple axis analog accelerometer with a range of +/-200 G. The X, Y, Z, ground and +3.3 V connections were soldered to the wiring harness. The ground and +3.3 V were supplied by the National Instruments USB-6211 DAQ. The Z axis and the X axis were used on each accelerometer due to the limited number of analog channels available. The Z axis was normal to the sample surface with +Z pointing towards the excavation equipment. The X axis ran parallel to the edge of the sample.

The accelerometers were mounted in enclosed 3D printed ABS sensor mounts. The ADXL337 was attached to the accelerometer mount by four screws. A cutaway was left in the base of the sensor mount for the wiring harness. The sensor mounts were pressed onto the rebar. To ensure a tight fit, the rebar was wrapped with electrical tape. The hexagonal tops of the sensor mounts were parallel to the surface of the panel. Seven sensor mounts were used during testing. These were arranged in a counterclockwise pattern around the sample starting with accelerometer 1 on the bottom right rebar end. There were three accelerometers on the right side of the sample, two on the top of the sample, and two on the left side of the sample. The sensor arrangement is shown in Figure 5.

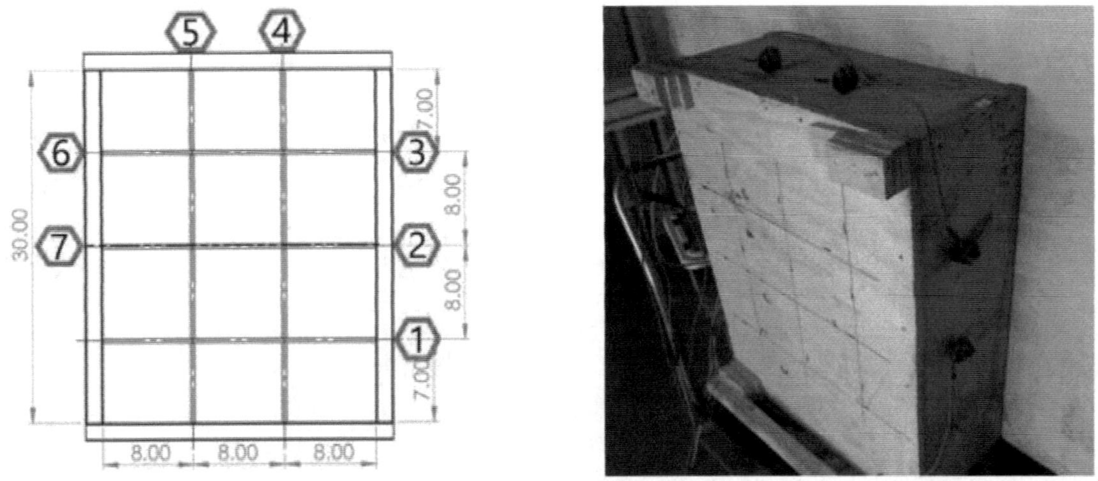

Figure 5. Accelerometer locations (left) accelerometers on actual panel (right)

A National Instruments USB-6211 DAQ was used to read the analog signals from the accelerometers. This DAQ supported sixteen analog channels. Fourteen of the channels were used for accelerometer data, one was used to monitor the accelerometer power supply voltage, and the remaining analog channel was not used. The specified maximum impact rate of the Bauer rotary hammer (1643E-B) was 3900 impacts per minute (65 HZ). Thermal and operating considerations lead to the hammer being operated at frequencies well below the maximum (below 50 HZ).

The sample rate of the DAQ was 100 HZ which translates to a recorded reading for each accelerometer in the X and Z direction every 0.01 seconds. This was sufficient to exceed the Nyquist frequency of the hammer at plausible operating conditions. Due to the recording sample rate, most cutting runs lasted approximately 210 seconds at which point recording stopped and the values for the readings per every 0.01 second were output in excel format. With this time constraint per run in the DAQ, each panel required multiple testing runs in order to excavate through the entirety of the shotcrete layer until reaching the shotcrete/concrete interface. For example, it took seven runs to reach the shotcrete/concrete interface in Panel 2, which was excavated using the impact hammer.

3. RESULTS

The following research results validate the hypothesis that waterjet removal of concrete/shotcrete liners will result in less collateral damage than conventional methods through comparison of visual inspection and processed GPR scans. For clarification purposes, the designed panels, along with their testing method, are shown in Table 3.

Table 3. Panel number and associated testing method

Panel	Testing Method
2	Impact Hammer
3	Impact Hammer
4	Impact Hammer
1	Waterjet
7	Waterjet
TEST	Waterjet

3.1 Visual Inspection Comparison of Kerfs and Damage of Steel Infrastructure

Visual inspection was performed after excavation was completed on each panel through analysis of the entirety of the excavation profile including the left, upper, right, and bottom walls, as well as the face where exposed aggregate was located. Figures outlining the inspection process of each excavation section can be found in Appendix C of Josef Bourgeois' dissertation [6]. Panels 2, 3, 4, 7, and TEST were examined at the EMI. Panel 1 was not visually inspected as it is currently being held at the SAI/RAMAX laboratory in California. Results of visual inspection measurements for all panels is shown in Table 4.

Table 4. Results of fracture/failure count and measured distance, measured delamination, and percent split aggregate (exposed)

Panel	Excavation Method	Fracture/Failure Count						Total Fracture / Failure Count	Total Measured Fracture / Failure Distance (mm)	Total Measured Delamination Distance (mm)	Percent Split Aggregate (Exposed)
		Left Wall	Upper Wall	Right Wall	Lower Wall	Excavation Face	Outside Excavation Profile				
2	Impact Hammer	3	1	3	2	1	3	13	66.5	0	62.7
3	Impact Hammer	7	3	6	4	0	2	22	165	131.5	62.5
4	Impact Hammer	4	3	2	0	0	2	11	57.5	92	80.7
7	Waterjet	1	0	0	0	0	1	2	19	0	0
TEST	Waterjet	1	1	1	0	0	1	4	23.5	0	-

3.2 Vibration Comparison from DAQ Results

During both forms of testing, acceleration in the Z direction typically exhibited higher magnitudes as a consequence of the Z direction being normal to the panel surface. Some of the higher magnitude acceleration values in the X direction during impact hammer testing could be contributed to when the angle of the chisel was changed from perpendicular to the excavation surface to having an attack angle in order to remove material in the X direction with relation to the panel face. Accelerometer results of waterjet testing generally appear as static noise with consistent vibration when excavating surfaces that do not have irregular geometry. Comparatively, results of impact hammer testing show frequent peaks in all accelerometers at values never once reached during waterjet testing, and thereby showing far less consistency with much higher vibration.

Positive and negative peak acceleration (g) values reached during testing on each panel are shown in Figure 6. Within this figure, the labeling of each panel that underwent excavation is preceded by the testing method used where "IH" stands for "Impact Hammer" and "WJ" stands for "Waterjet". For Panel 7, acceleration readings during waterjet excavation never reached above 6 g. With Panel 1, there were instances when accelerations between 10 and 20 g were recorded when the jet stalled at the end of a traverse before reversing direction or when the nozzle traversed across areas of kerf geometry where additional free surfaces existed because of adjacent kerfs. The longest period of time that the nozzle assembly stalled at the end of a traverse was approximately 15 seconds. This could be avoided in future applications by utilizing hydraulics to traverse the nozzle assembly rather than air compression. Overall, even with these relatively infrequent spikes in recorded acceleration, values never exceeded 20 g during waterjet testing.

A section of Panel 4 was excavated using hand hammering with the chisel and a sledge hammer, instead of using the Bauer tool, to draw a comparison of different impact techniques. During hand hammering, acceleration peaks between 17-72.5 g were recorded which were comparatively lower than typical peaks reached with the Bauer tool however, still higher than that of the waterjet system. The higher vibrations induced through impact hammer testing are likely the cause of delamination at the interface of the shotcrete/concrete layer as noted in Table 4.

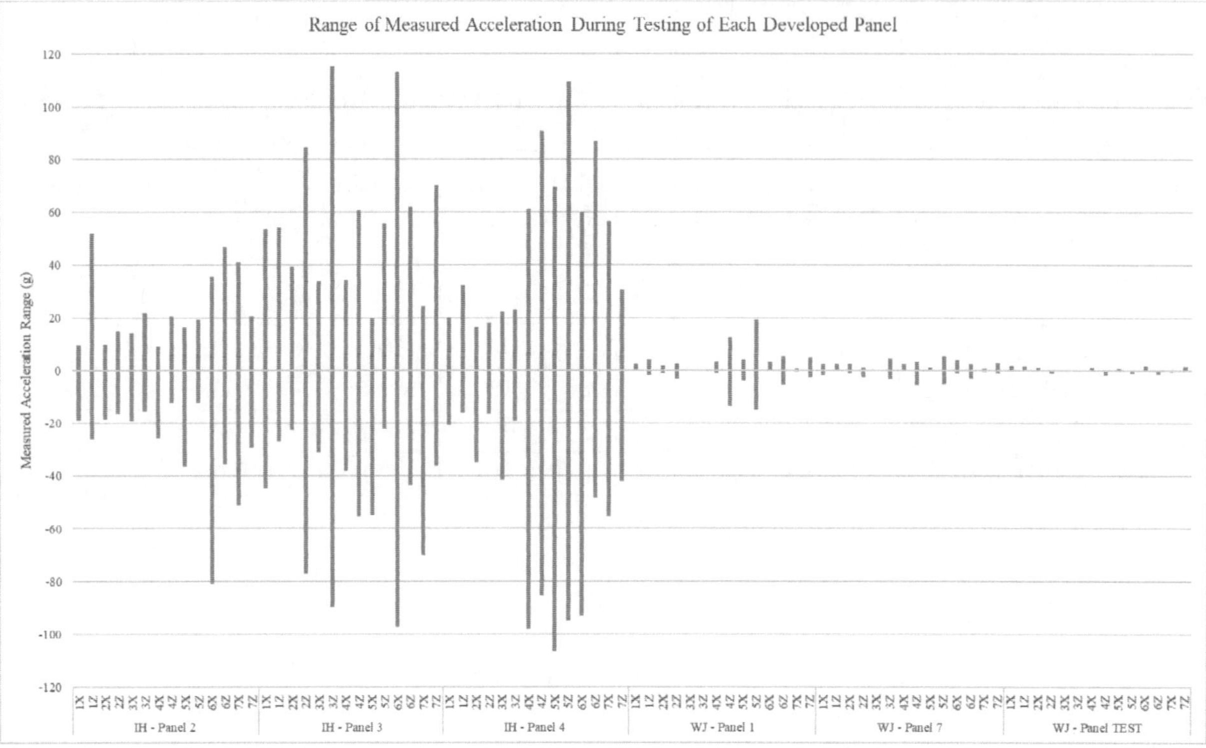

Figure 6. Range of measured acceleration during testing of each developed panel

4. DISCUSSION

This section summarizes completion of the research objectives through empirical comparison of excavation methods, as well as an analysis of the proposed hydro-excavation technology.

4.1 Visual Inspection

Visual inspection of Panels 2, 3, 4, 7, and TEST exemplified stark differences in the occurrence of fractures/failures, delamination, damage to steel wire mesh, and the incidence of split aggregate associated with each testing method. As previously stated, Panel 1 was not visually inspected as it is currently being held at the SAI/RAMAX laboratory in California. Table 4 shows that impact hammer excavation caused more counts of fracture/failure, compared to that of waterjet cutting, along the excavation profile and surrounding shotcrete layer with 13 (66.5 mm), 22 (165 mm), and 11 (57.5 mm) for Panels 2, 3, and 4 respectively. Panels 7 and TEST resulted in fracture/failure counts of 2 (19 mm) and 4 (23.5 mm). While the panels tested by the waterjet system did result in measured fractures/failures, there was no visual indication of delamination, damage to wire mesh, or split exposed aggregate. This is especially important because the panels tested using the impact hammer contained damaged wire mesh, measured delamination in two of the three panels, and measured over 50% split exposed aggregate for all panels. The highest measured delamination within the excavation profile of a panel subjected to impact hammer testing was 131.5 mm from Panel 3, which also had the highest recorded vibration from accelerometers.

4.2 Vibration Data Acquisition

Accelerometer data gathered during testing of Panels 2, 3, 4, 1, 7, and TEST illustrated significant difference in vibrations induced by the chosen excavation tools with the impact hammer consistently recording acceleration far higher than that of the waterjet. This data clearly demonstrates that use of a high-pressure waterjet systems will translate substantially less vibrations to the embedded steel reinforcement. The high readings of acceleration shown in Panel 1 can be reduced through adjustment of operating parameters, specifically avoidance of stalling the nozzle assembly while keeping a consistent traverse speed and avoiding standoff distances of less than 8.9 centimeters (3.5 inches). It's important to note that, while particularly high acceleration was recorded for Panel 3, experienced use of the impact hammer could result in lower readings like that of Panel 2, however, these are still considerably higher than data gathered for all waterjet panels.

5. CONCLUSIONS

The results obtained from visual inspection, and collected accelerometer data correlate with observations made in the literature favoring the use of hydrodemolition in bridge deck removal and repair work [4], specifically those identifying structural benefits to its use including:

- Providing a rough surface profile for excellent mechanical bond to repair materials;
- Eliminating surface micro-fracturing;
- Not fracturing exposed aggregates;
- Minimizing vibration; and
- Cleaning steel reinforcement and leaving it undamaged.

The data collected for this research project provides supplemental validation of the benefits of hydrodemolition. According to Asadi et al, the Federal Highway Administration reported that 11% of the bridges in the United States are rated as structurally deficient and over 30% of existing bridges have exceeded their 50-year design life [7]. This means that bridge deck analysis and rehabilitation work will substantially increase in the years to come. The findings from this research, specifically through the reduction of vibration to embedded steel reinforcement and avoidance of cutting the aggregate within the concrete layer for an improved bonding surface to the new overlay material, can be used to further advance the knowledge and use of high-pressure waterjet systems for bridge deck rehabilitation. The next step of this research agenda would be to adapt the waterjet system used from this study to an existing tunnel or mine site, where large sections of an existing support liner could be excavated as part of a full-scale pilot test. Based on the vibration analysis conducted in this study, accelerometers could be embedded into sections of support liner for continued collection of data during future empiric testing. Additionally, to enhance visual inspection of the excavation profile developed in a shotcrete layer, liquid penetrant inspection (LPI) could also be utilized for detection of impact failure/fracture.

Finally, the findings from this study could also be used to optimize the jet's specific energy for fragmentation. Through this process, the time required to remove a volume of shotcrete could be characterized in hopes of increasing the excavation rate as compared to conventional mechanical methods. For commercial application, a cost-benefit and safety analysis between waterjet systems and traditional tools, such as impact hammers, would be beneficial for advancing the commercial relevance of the technology in specific industrial applications.

6. ACKNOWLEDGEMENTS

The research was made possible by the U.S. Department of Transportation (DOT) for financial support provided under Grant No. 69A3551747118 and administered by UTC-UTI at CSM. The authors would also like to acknowledge the support and contributions provided by Dr. Fun-Den Wang, Dr. Eugene Nebeker, Mr. Roy Bowling, Dr. John Steele, Mr. Brian Asbury, Mr. Brent Duncan, Mr. Robert Pike, Dr. Jamal Rostami, Dr. Marte Gutierrez, and Dr. Rennie Kaunda. The support of the many professionals at the CSM Energy, Mining and Construction Industry Safety Program (EMCIS) and the EMI were instrumental in facilitating this research and was also greatly appreciated.

7. REFERENCES

[1] L. Malmgren, E. Nordlund, and S. Rolund, "Adhesion strength and shrinkage of shotcrete," Tunn. Undergr. Sp. Technol., vol. 20, no. 1, pp. 33–48, Jan. 2005.

[2] E. C. Poeck "A Performance Evaluation of Various Nozzle Designs for Waterjet Scaling in Underground Excavations," M.S. thesis, Colorado School of Mines, Golden, CO, 2008.

[3] L. Bodnarova, L. Sitek, J. Foldyna, T. Jarolim, and R. Hela, "Application of water jet technology for concrete repair," MM Sci. J., vol. 2018, no. June, pp. 2385–2389, 2018.

[4] J. Wenzlick, A. Careaga, and S. Clark, "Guide for the Preparation of Concrete Surfaces for Repair Using Hydrodemolition Methods," International Concrete Repair Institute, Des Plaines, IL, Tech. Report, FM 7.10 Exhibit 1, Sep. 2004.

[5] J. D. Wenzlick, "Hydrodemolition and Repair of Bridge Decks," Missouri Department of Transportation Research, Development and Technology, Jefferson City, MO, Tech. Report. R197-025, 16 Dec. 2002.

[6] J. Bourgeois, "Characterizing the Excavation Process Associated with Removal of Shotcrete Reinforced Underground Structural Liners Using Waterjet Technology," dissertation, Colorado School of Mines, Golden, CO, 2020.

[7] P. Asadi, M. Gindy, and M. Alvarez, "A Machine Learning Based Approach for Automatic Rebar Detection and Quantification of Deterioration in Concrete Bridge Deck Ground Penetrating Radar B-scan Images," KSCE J. Civ. Eng., vol. 23, pp. 2618–2627, 2019.

2022 WJTA Conference and Expo
November 2-3 • New Orleans, Louisiana

Paper

Erosion of an Air-injected Plain Waterjet (iPWJ)

J. Poloprudský
Institute of Physics of Materials, CAS,
Brno, Czech Republic

M. Schüler, T. Bergs
Fraunhofer Institute for Production Technology IPT
Aachen, Germany

ABSTRACT

The article describes the fundamental characteristics of a plain waterjet (PWJ) tool enhanced by air. An injection-type abrasive waterjet (AWJ) head was used, in which the entrained abrasive was replaced by a defined air input. A two-phase PWJ was created by the injection principle (iPWJ). In dependence on the air pressure supplied, the footprint of the perceived jet tool was characterized. A standard aluminum alloy EN AW-5083 was chosen for the material to be machined. The experiments were designed with two water pressure levels, and four pressure conditions of the air inlet that varied from blocked air, atmospheric conditions, and overpressure. An evaluation was conducted for the visual shape of the free jet and the perceived erosion characteristics of the specimen by scanning electron microscopy (SEM).

Organized and Sponsored by WJTA®

1. Introduction

The demand for improvements in the application of waterjet machining has increased in recent years. Plain Waterjet (PWJ) machining still offers some advantages for special applications over its abrasive-enhanced counterpart [1]. However, both the effectiveness of the PWJ and the quality of the surfaces machined were of great significance in recent years. A major route was the optimization of the jet tool by modulating or pulsating waterjet [2], high-pressure pumps, or cryogenic jets [3] to overcome the use of abrasives. High effort on the system side and the related costs of the machine systems are common problems for all approaches. Therefore, a high-adaptive and more accessible approach was proposed to encourage industrial acceptance and meet the needs of small and medium enterprises [4]. A simple but innovative approach to changing the erosion characteristics of a PWJ is presented here by adapting a standard abrasive waterjet (AWJ) tool to be used with pressurized air instead of entraining abrasives.

2. State of the Art

PWJ, as the systematically simplest form of a waterjet, just squeezes pressurized water through an orifice. If solid particles or abrasives are added to the waterjet, this is generally referred to as an abrasive waterjet. Abrasive waterjet variants can be subdivided according to the type and location of the abrasive supply. If the abrasive is fed through an injector system, this standard in the industry is referred to as an injection-type Abrasive Waterjet (AWJ), which is the system currently predominantly used in manufacturing. Opposite, a previously mixed suspension of water and solids is referred to as an Abrasive Suspension Jet (ASJ), which is currently used for special applications or reconstruction but has not been established in production scenarios yet.

The choice of the waterjet mode is typically given by its application. PWJ is typically used for the processing of relatively soft or sensitive materials. AWJ enables the processing of virtually all materials and represents an extension of the PWJ on the system side by adding a mixing unit to the device. Due to the Venturi effect, the high velocity of the PWJ creates a local vacuum in the mixing chamber, which sucks an airstream that is used to transport the added particles and inject them into the mixing chamber. The kinetic energy of the pure waterjet is then used to accelerate the added particles and to shape the composing jet through the focus tube. The resulting jet tool consists of water, abrasive, and air that all expand to fill the inner diameter of the focus tube. This is contrary to the ASJ where both water and abrasive are accelerated simultaneously through a nozzle, typically giving better efficiency. The construction of an ASJ tool is very similar to that of a PWJ, with the nozzle as the essential wear part. Categorization can be done based on the phase composition of waterjet variants (**Figure 1**):

- one-phase PWJ (water)
- two-phase ASJ (water + abrasives)
- two-phase iPWJ (water + air)
- three-phase AWJ (water + abrasives + air)

A PWJ, containing just water, consists of a single-phase fluid. A two-phase abrasive jet can be generated by added solids (ASJ). The absence of an air phase is promising for higher cutting performance with better efficiency and intensity. [5]. A newly introduced two-phase injection plain waterjet (iPWJ) is based on a regular injection AWJ head without adding abrasives to present an air-enhanced PWJ tool. As a note, the aim can be considered contrary to waterjet cutting, where kinetic intensity is key. However, the industrial standard for abrasive waterjet cutting is still a three-phase AWJ [6].

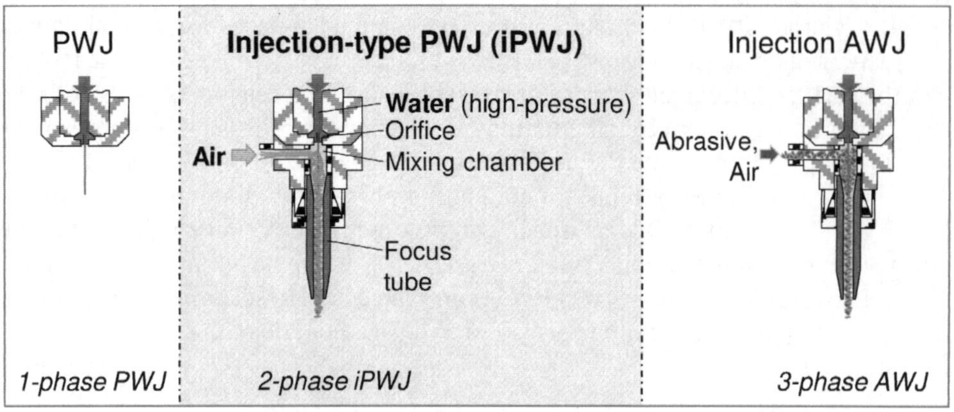

Figure 1. Schematic of an iPWJ configuration

When it comes to AWJ, the acceleration of the abrasive particles by the initial PWJ down the focus tube leads to subsequent surface contacts and wear, accompanied by a lower overall velocity. Due to interactions with the high air content within the jet, a turbulent waterjet is created. According to the injection principle, the main energy transfer takes place behind the mixing chamber. Here, the inner surfaces of the focus tube are intensively stressed during an unsteady acceleration process. The expansion of the jet, the presence of a considerable amount of air, and the overall lower efficiency of this mixing process must be compensated. For example, a higher system pressure leads to increased wear of the orifice or other high-pressure components and the pump itself. In earlier studies, the gaseous and compressible phase of entrained air was neglected. It should be noted that an injection AWJ consists of up to 95 % volume of air. Later studies considered this complex phenomenon fundamentally [7-8]. Several methods were considered, such as finite element simulation, jet visualization, high-speed imaging, pressure profile parameters, or empiric studies [9]. However, the body of research mainly consists of parametric studies combined with jet visualization.

Huang et al. [10] modeled a velocity field inside the injection head using computational fluid dynamics (CFD). The results showed that the highest waterjet velocity was achieved directly behind the entrance into the mixing chamber and slowly decelerates with increasing distance from it, developing a regular flow pattern. Osman et al. [7] analyzed a two-phase flow in a focus tube of a typical diameter under varying waterjet velocities, as well as varying nozzle and mixing tube diameter ratios. The airflow in the study was either restricted completely by blocking the air inlet or allowing free atmospheric flow. The analysis was approached in two main steps. The first step was a static visualization of the waterjet structure at the focus tube entry and downstream towards the nozzle. The second step of the analysis was based on static pressure measurements down the focus tube. In the case of the free atmospheric airflow, the pressure distribution along the focus tube follows a similar trend for all water velocity levels only differing in the resulting jet velocity and proportions of the pressure-affected areas. The first observed area in the focusing tube was the expansion region where a significant decrease in local pressure was observed. The negative pressure gradient in this region was related to air acceleration. The second area inside the focus tube was considered the region of constant pressure. In this region, the airflow reached a roughly uniform state. The third area, called a recompression region, was where the local pressure rapidly changes to the atmospheric condition. The size of this area was dependent on the jet velocity v_{Jet}. For lower jet velocities $v_{Jet} < 447\ \mathrm{m/s}$, the recompression zone was spread out over the jet length. In the case of high velocities $v_{Jet} > 447\ \mathrm{m/s}$ the recompression region was located at the nozzle. In the case of blocked air suction, the recompression zone for low jet velocities and small jet diameters occurred gradually along the length of the focus tube. For blocked air suction, in combination with high jet velocities and large jet diameters, the pressure development in the focus tube consists of two areas. The first region was of constant pressure, which approaches vapor saturation pressure. The second region was a recompression region with a high-pressure gradient that was concentrated close to the focus tube exit [7].

Himmelreich [11] divided the phenomena inside the focus tube into two categories based on the ratio of the inner diameter of the focus tube to the inner diameter of the orifice ($\emptyset d_F / \emptyset d_O$). In the case of $\emptyset d_F / \emptyset d_O = 2.75$, Himmelreich observed pipe flow in the focus tube. A free waterjet was observed in the focus tube with a higher ratio of $\emptyset d_F / \emptyset d_O = 4$. Transferred, this means that for small diameter ratios $\emptyset d_F / \emptyset d_O$ the air was compressed while in larger ratios $\emptyset d_F / \emptyset d_O$ the entrained air was accelerated by the jet.

Momber [9] focused on the processes inside the mixing chamber approaching from an energetic point. An energy transfer from the high-speed water to the solid particles was described, which are introduced into the mixing chamber at a relatively low speed. Due to the lower pressure inside the chamber, the air was sucked into the mixing chamber. During this energy transfer process of the involved phases, the primary waterjet decelerates to generate the exiting multi-phase jet of water, abrasive, and air combined.

In the recent work of Bergs et al. [12], it was seen that the interaction of various phases occurs at several levels. The injection principle introduces some phenomena that alter the machining results. The researchers primarily compared the resulting surfaces of quenched and annealed 42CrMo4 either with AWJ or ASJ. Both were then post-treated using a two-phase iPWJ. The Interaction of the involved jet phases revealed fundamental effects of the gaseous phase within the machining results. Nevertheless, the proceedings of their work [13] proposed a great potential for multi-stage machining.

Even though the mass flow of the sucked air is very small compared to the mass flow of the other phases of an AWJ, the influence of the air phase and its comparative huge volume has been neglected in the past. However, the role of the air phase and its importance for process understanding is still lacking. The volume of entrained air has significant effects on the shape and stability of the jet when it exits the tool head [7].

3. Methodology

The effect of entrained air on the characteristics of an iPWJ tool and its effect on material erosion were not investigated yet. Consequently, this study uses a modified iPWJ configuration and enhances inlet conditions. Here, the inlet port was hermetically modified to create different conditions, for example by pressurization. The experiment concentrated on the erosion characteristics of the modified iPWJ tool onto an aluminum specimen. By varying both the water pressure (p_W) and the pressure of supplied air (p_A), the ratio of the phase composition, and thus its effect on the waterjet tool, were investigated. A comparison was made between the footprint at a water pressure level of $p_W = 200$ MPa and $p_W = 400$ MPa, respectively. Accordingly, the airflow into the mixing chamber was controlled at the abrasive inlet. The air pressure ranged from a blocked inlet p_A (0), through atmospheric conditions at the inlet p_A (0) = 1000 hPa up to pressurization of p_A (2) = 2000 hPa and p_A (3) = 3000 hPa, respectively. The main variable to be observed was the airflow introduced into the mixing chamber. Here, the airflow was set by a mechanical pressure regulator and checked with a mechanical flow meter (**Figure 2**). A preliminary study was conducted to choose a constant feed rate $v_F = 2000\ mm/min$ in combination with a standoff distance (SOD) between the surface of the specimen and the nozzle of the focus tube at two levels: SOD = 4 mm and SOD = 54 mm. The jets with given parametric variations according to **Table 1** were then applied to the aluminum specimen. The footprints were then studied visually to understand the effect of air conditions on the tool characteristics by its shape and erosion effects.

Table 1. Factors and levels of the experiments

Material of the specimen	Water Pressure p_W [MPa]	Standoff Distance SOD [mm]	Air Pressure p_A [MPa]	Feed rate v_F [mm/s]	Orifice Diameter $Ød_O$ [mm]	Focus Tube Diameter $Ød_F$ [mm]
AW 5083	200 - 400	4; 54	0 - 0.3	2000	0.20	0.76

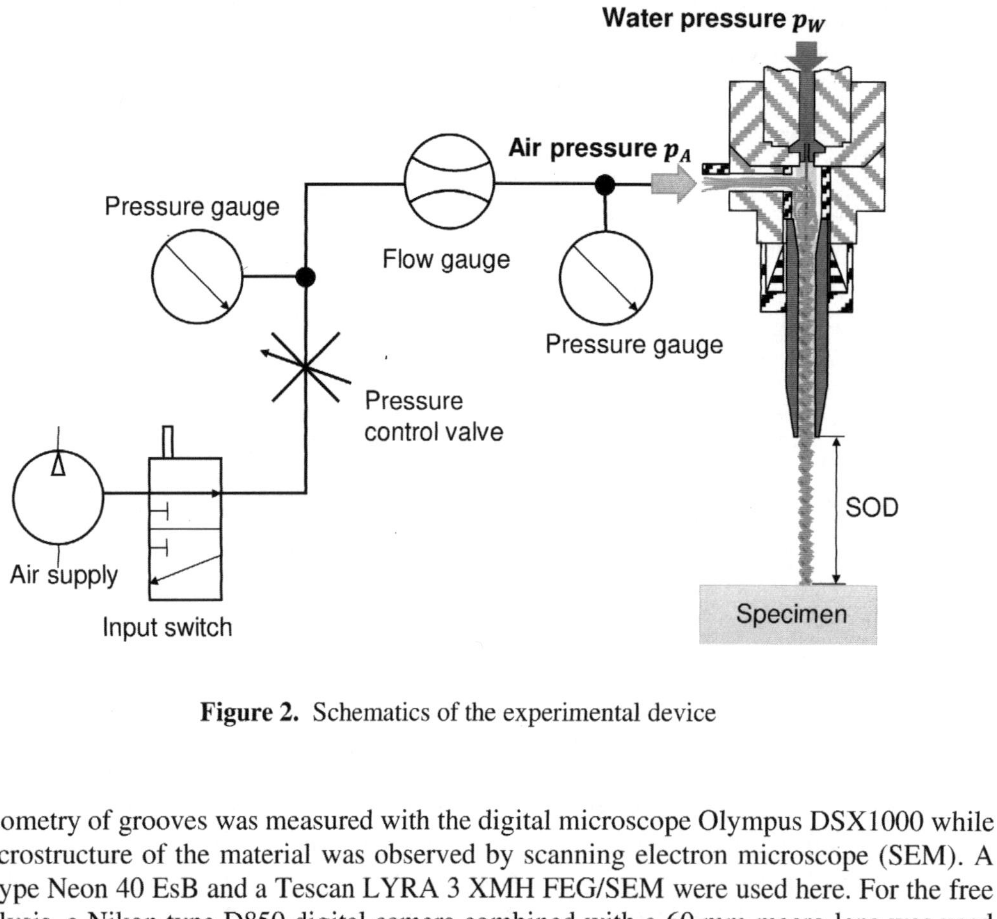

Figure 2. Schematics of the experimental device

The geometry of grooves was measured with the digital microscope Olympus DSX1000 while the microstructure of the material was observed by scanning electron microscope (SEM). A Zeiss type Neon 40 EsB and a Tescan LYRA 3 XMH FEG/SEM were used here. For the free jet analysis, a Nikon type D850 digital camera combined with a 60 mm macro lens was used and the images were processed with ImageJ software.

The experiments were carried out on a customized 5-axis waterjet machine H.G. Ridder type E, equipped with a pressure intensifier pump capable of pressures up to 400 MPa. A standard injection head was used for the experiments, equipped with a 0.20 mm orifice in combination with a 0.76 mm focus tube. The abrasive inlet was used to control the air-flow input. The airflow at different levels of water pressure p_W were measured for different levels of air pressure p_A as shown in **Figure 3** a). For comparison, the suction pressure of a blocked inlet into the mixing chamber over the applied levels of water pressure p_W is shown in **Figure 3** b).

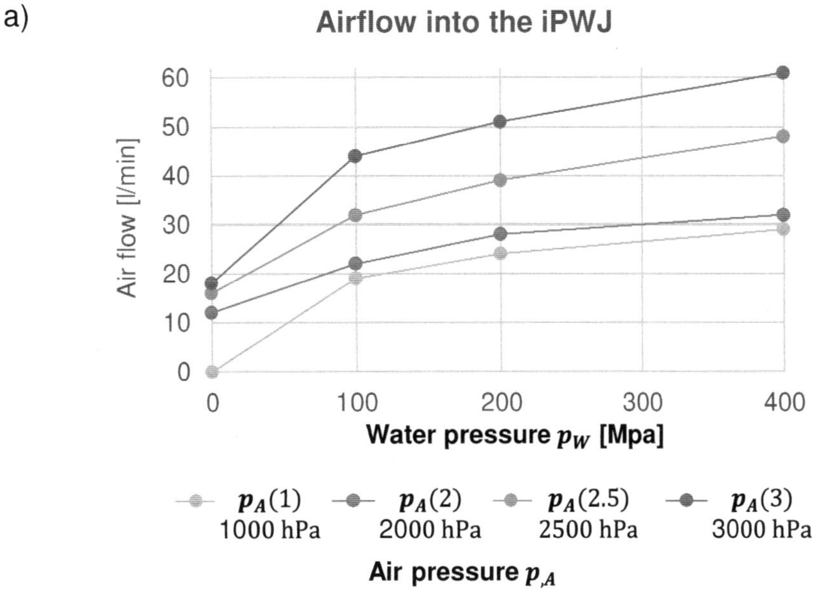

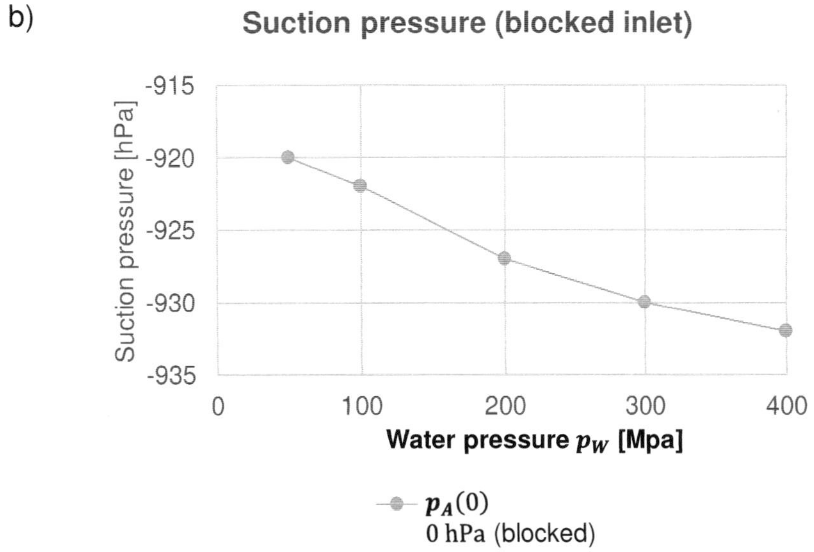

Figure 3. a) Measured airflow $\dot{V}_A$ over the waterjet pressure p_W at several air pressure levels p_A b) Suction pressure inside the mixing chamber for the investigated levels of water pressure p_W.

A wrought aluminum alloy EN AW-5083 with magnesium and manganese as the basic alloying elements was polished and prepared as a specimen. **Table 2** gives the composition measured by Energy dispersion spectroscopy (EDS). The main constituent particles observed in material matrices are MnFe and MgSi, as documented in **Figure 4**, elongated and uniformly distributed in the material.

Table 2. Alloy composition as measured by EDS

Element	Al	Mg	Mn	Fe	Si
Weight %	94.19	5.07	0.5	0.14	0.11

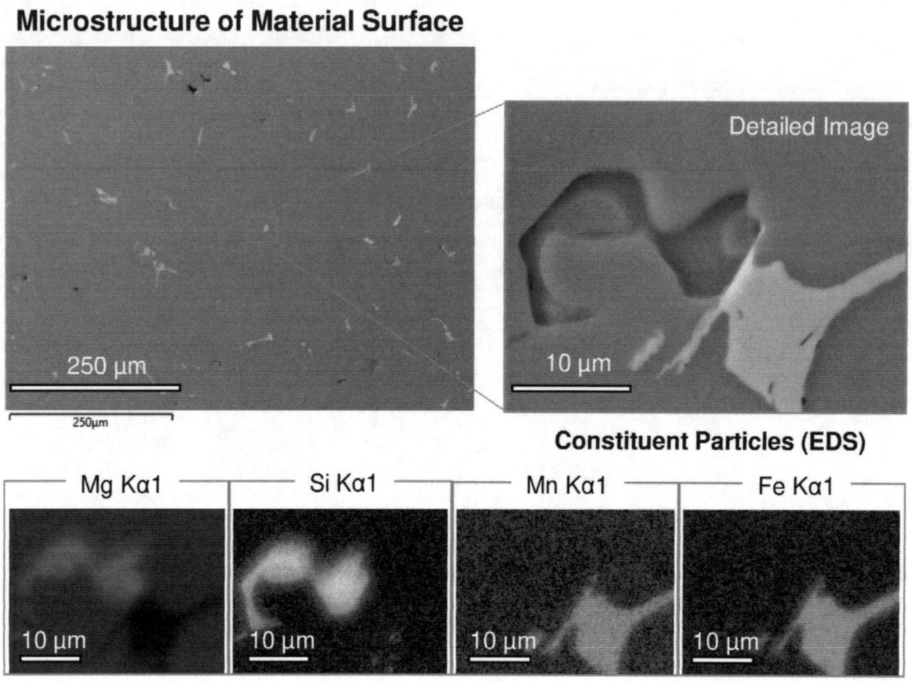

Figure 4. Microstructure of experimental material and EDS analysis of the constituent particles

4. Results and Discussion

Jet shape

In the first part, the airflow for different levels of both air pressure p_A and water input pressure p_W was measured. Then the visualization of the resulting shape of the free jet was presented. In the second part, the dimensions of the footprints at the bottom and side regions on the aluminum specimen were measured using SEM. The objective of the observation was to characterize the erosion pits with a change in both air pressure p_A and SOD. **Figure 5** compares the images taken at four different levels of air pressure p_A. The images of the free air iPWJ were taken under identical exposure conditions. The photographic images were then processed by applying a contrast threshold to enhance the visibility of the core of the jet and the jet disruption. Here, the jet shape shows a stable widening with increasing SOD in the case of air pressure ranging from the blocked inlet $p_A\,(0)$ to $p_A\,(2) = 2000\,hPa$. At $p_A\,(3) = 3000\,hPa$, the jet shows a one-sided disturbance in width directly after the focus tube. The direction of this irregularity is most likely based on the position of the air inlet.

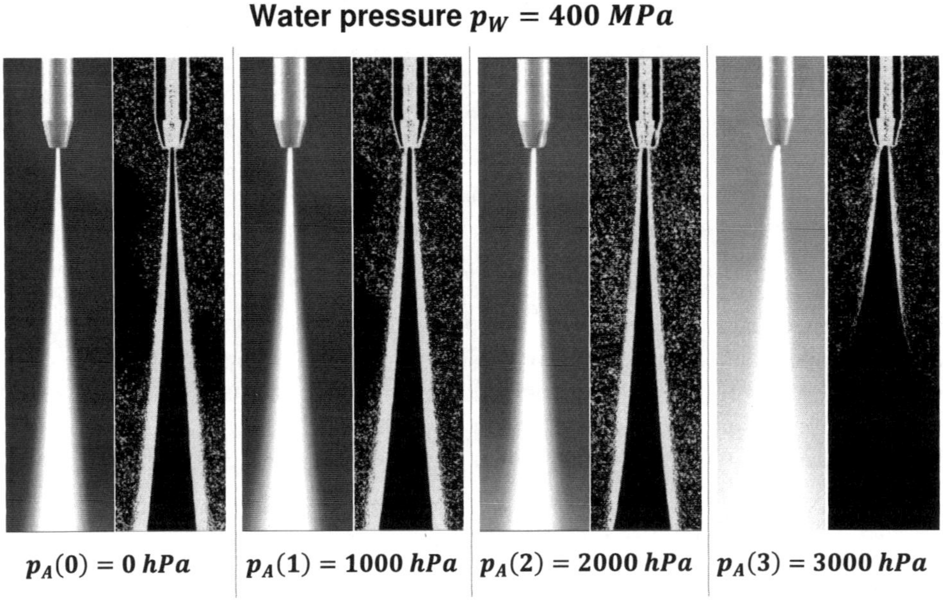

Figure 5. Images of the jet shape for a fixed water pressure level $p_W = 400\ MPa$ over varying air pressure levels p_A. Each photograph was image processed to highlight the jet shape.

Surface morphology SEM analysis

Figure 6. Optical microscopy schematic overview of the samples for SEM

Figure 6 gives a visual overview of the machined specimen. In the naming, the water pressure p_W is noted in MPa, the number after p_A represents the air pressure condition, accompanied by the SOD in mm. A visual comparison revealed significant differences in the groove width in dependence on the air situation. This is especially visible in the case of SOD = 54 mm together with a water pressure p_W = 400 MPa. However, the water pressure of 200 MPa shows a similar trend here.

In **Figure 7**, an overview by SEM imaging is given for a high level of water pressure p_A = 400 MPa. At the low level of SOD = 4 mm, similar groove widths for all four levels of p_A were evident. Opposite to the high level of SOD = 54 mm, a difference was visible in the width of the grooves. The footprint with a blocked abrasive inlet p_A (0) was wider compared to the one created with an open inlet to atmospheric pressure p_A (1). Further, the footprint created with an air pressure p_A (2) = 2000 hPa was similar in appearance to that of atmospheric pressure. However, the centerline of the groove appeared more serrated. An increase in effective jet diameter was enforced at the highest pressure level p_A (3) = 3000 hPa. The center line of the groove created here appeared most uneven and serrated. As an interim conclusion, the observations at a pressure level p_W = 400 MPa revealed, that the jet footprint was at its widest dimension when the air inlet was blocked. For an open inlet, the groove narrowed first until an overpressure of supplied air forced the jet to widen again.

Figure 8 shows an equivalent comparison of grooves created by the low level of water pressure p_W = 200 MPa. Like the previous results, a widening of the jet footprint based on the air pressure level was observed here. Further, the impact of air pressure on the jet's effectiveness was observed very clearly. For both levels of SOD, the jet footprint was most uniform for p_A (1) and p_A (2). Note that the low level of SOD leads to grooves that appear more uniform at p_A (2). In contrast, the high level of SOD revealed a more uniform jet footprint at atmospheric pressure p_A (1). Both the blocked inlet p_A (0) and the pressurized inlet p_A (2) created uneven erosion pits. The high level of SOD supports smaller and more isolated erosion pits at both p_A (0) and p_A (3), respectively.

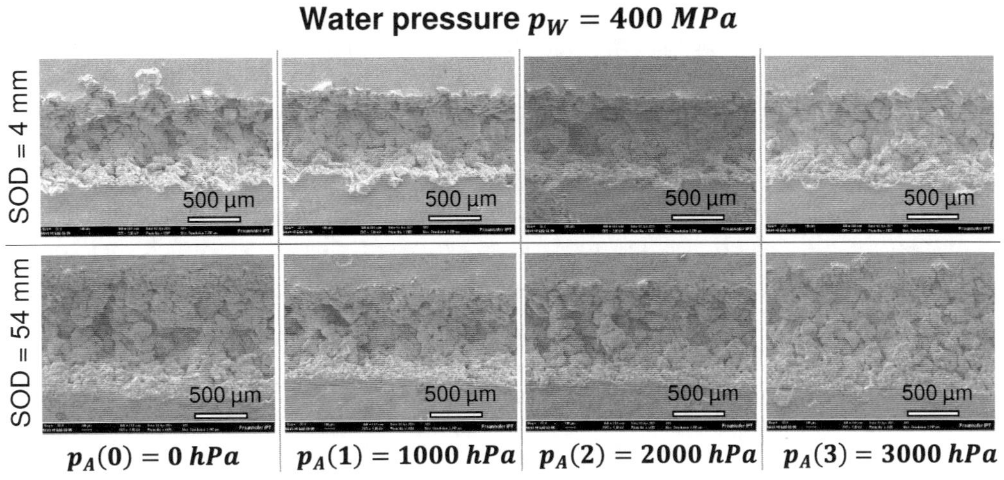

Figure 7. SEM overview of erosion groves machined at a high level of water pressure p_W = 400 MPa: Both levels of SOD at all levels of air pressure p_A

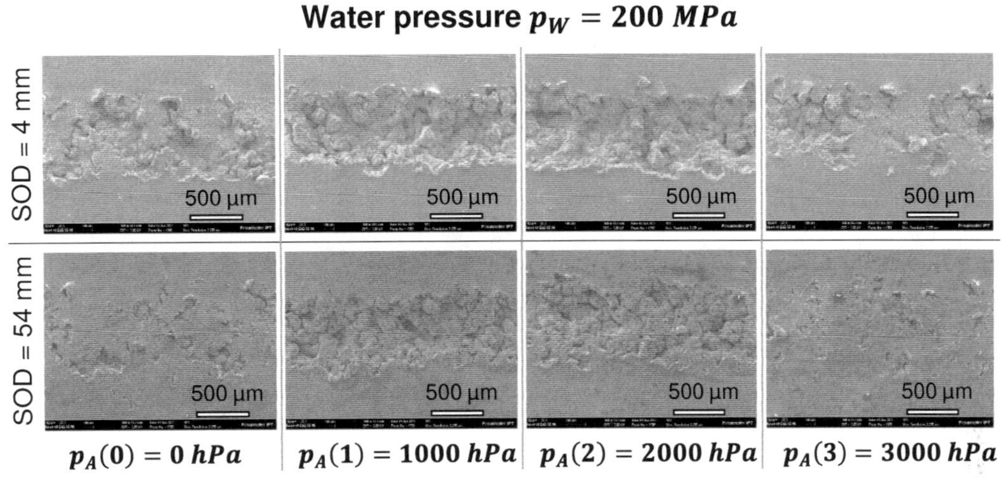

Figure 8. SEM overview of erosion groves machined at the low level of water pressure p_W = 200 MPa: Both levels of SOD at all levels of air pressure p_A

Figure 9 and **Figure 10** compare the detailed images for both levels of SOD at a lower level of water pressure p_W = 200 MPa. Since the changes in jet erosion were more visible on SEM, a detailed overview was given for both the central region and edge region over the levels of air pressure. **Figure 9** shows footprints created at a lower level of SOD = 4 mm while **Figure 10** is related to the high level of SOD = 54 mm, respectively.

In **Figure 9**, the central line at p_A (0) revealed separated pits of deep erosion. The area between the pits was covered with small cavities which can be seen in **Figure 9 a)**. The Pits were characterized by narrow edges and smooth bottoms. A central region of the footprint related to p_A (1) was shown in **Figure 9 b)**. This area consists of a homogenous erosion trench. Compared to pA (2) in **Figure 9 c)**, the area shows more smooth facets. Here, the details also showed several erosion pits of shallow expression and some deeper cavities. It appears that the smooth facets were attributed to both the plastic deformation and the shear of the material caused by lateral outflow. The footprint related to p_A (3) in **Figure 9 d)** shows similar structures to the one corresponding to p_A (0). It features shallow erosion cavities on the surface and marks of plastic deformation between the larger erosion pits, which were spread across several grains. At the edge region of all footprints, there were small erosion cavities. Both details in **Figure 9 f)** and **g)** show a higher slope of the groove compared to **Figure 9 e)** and **h)**. The edges of the grooves eroded with p_A (2) and p_A (3) in **Figure 9 f)** and **g)** respectively also showed a great number of small but deep cavities. These could be caused by lateral flow and related to the groove gradient. The crack propagating in **Figure 9 e)** shows that the cracks grow on the edges of erosion pits or inside them. **Figure 9 h)** visualizes that constituent particles, which were abundant within the material, appeared to be preferred places for crack initiation and propagation.

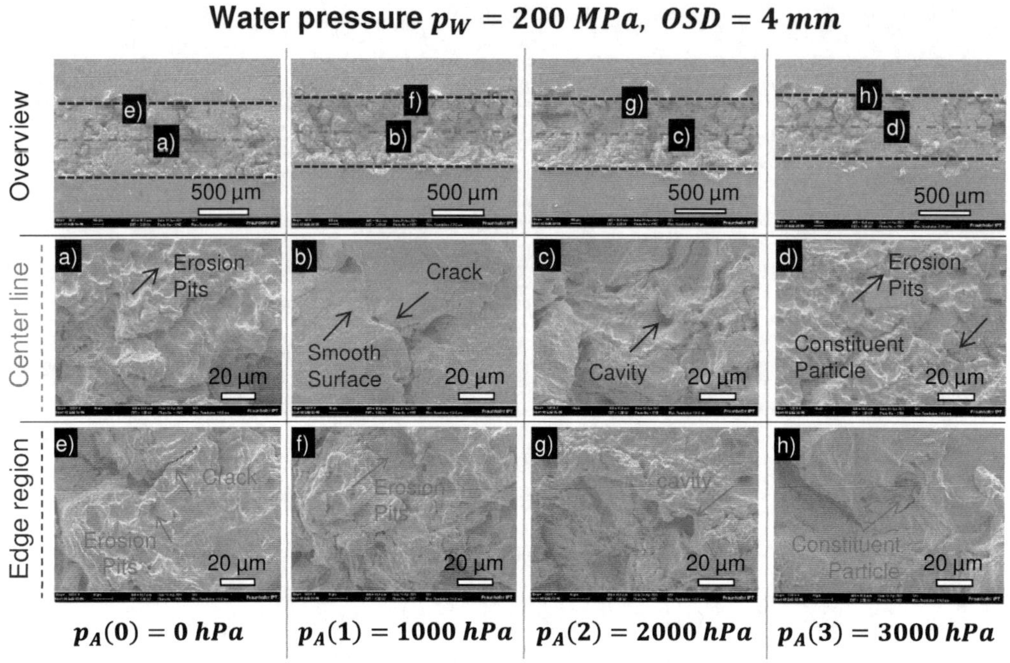

Figure 9. Detailed comparison of footprints machined with the low level of water pressure p_W = 200 MPa at the low level of SOD = 4 mm over variation of the air pressure p_A.

In **Figure 10** a higher level of SOD = 54 mm was displayed, while the basic layout remains identical to the previous Figure 9. In the results, a notable difference was seen in the central region, particularly in the width of pits related to the outer air pressure levels p_A (0) and p_A (3), respectively. In both the detailed **Figure 10 a)** and **d)**, the number of shallow erosion cavities increased, but with a decrease in its diameter, most notably seen in **Figure 10 a)**. In comparison, the footprints related to the remaining p_A (1) and p_A (2) were shallower than their counterparts at the lower level of SOD = 4 mm in the previous **Figure 9**. This translates to a less serrated bottom in the region around the center line of both **Figure 10 b)** and **c),** where the bottom of the groove showed a very linear array of pits that were created at p_A (1) in **Figure 10 b)**. A similar linear array of pits can be seen on the bottom of the groove related to p_A (2) in **Figure 10 c)**. The separated erosion pits at the center line were observable in **Figure 10 d)** at both air pressure levels p_A (0) and p_A (3). A steep drop between the outer wall of the groove and the unaffected surface of the workpiece was recognized. However, the edge region was comparable to the result of SOD = 4 mm. **Figure 10 e)** shows a similar characteristic. It should be mentioned that the edge also showed several cracks near the cavity entrance. For both levels p_A (1) and p_A (2), a steep curve at the edge and considerable pileups were seen in **Figure 10 g)** and **h)**, respectively. Pileups were considered in protecting the surface behind them from lateral jetting. **Figure 10 f)** shows a crack propagating directly from the pile-up. **Figure 10 h)** reveals a high gradient at the groove edges and several deep cavities. The material behind the pileup appeared rather flat and smooth.

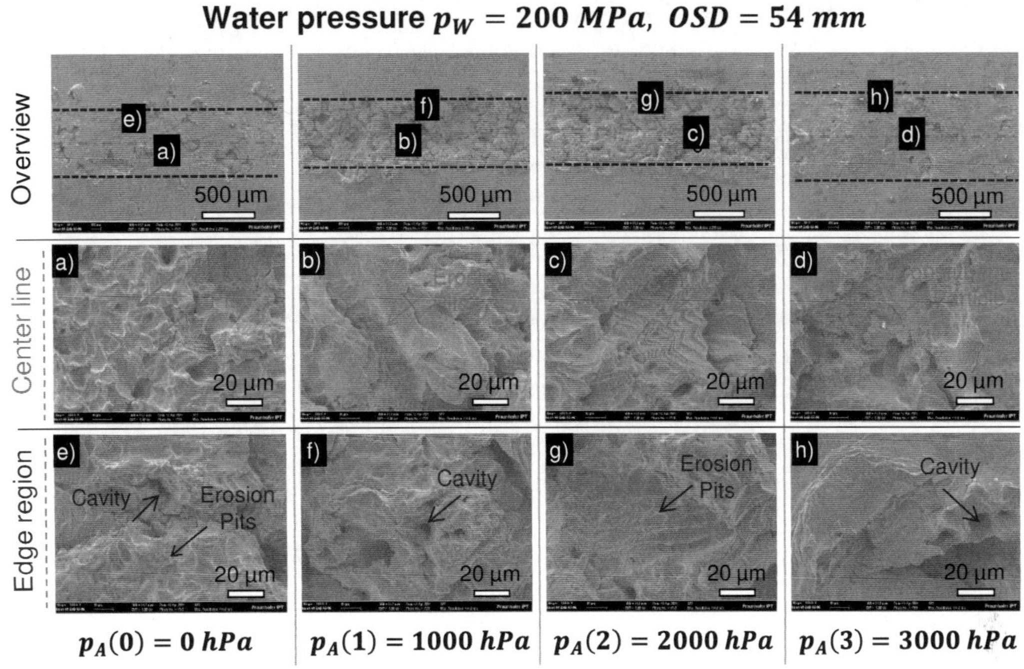

Figure 10. Detailed comparison of footprints machined with the low level of water pressure $p_W = 200$ MPa at the high level of SOD = 54 mm over variation of the air pressure p_A.

The derived diameter from the jet footprint was plotted in **Figure 11**. Here, the average diameter of the jet tool was obtained from the SEM images by ten individual measurements. For the low level of SOD = 4 mm, a variation in the air pressure revealed only a minor effect on the derived diameter of the jet. More important appeared the influence of the water pressure p_W. The trend was similar, but more pronounced for the high level of SOD = 54 mm. Although the jet diameter could be shifted following the applied levels of air pressure, it was significantly increased at the high level of SOD for both levels of water pressure. There the entrained air helps to disintegrate the jet under free-air conditions, similarly to the creation of cavitation bubbles on the other end. A local minimum of the groove width was found at a moderate air pressure level between p_A (1) and p_A (2). The shape of the trend is convex here. Note that due to the discontinuity of the footprints, some deviations occurred. Notably hard to measure is the location of the groove width at the low level of water pressure $p_W = 200$ MPa in combination with the outer levels of air pressure p_A (0) and p_A (3), respectively.

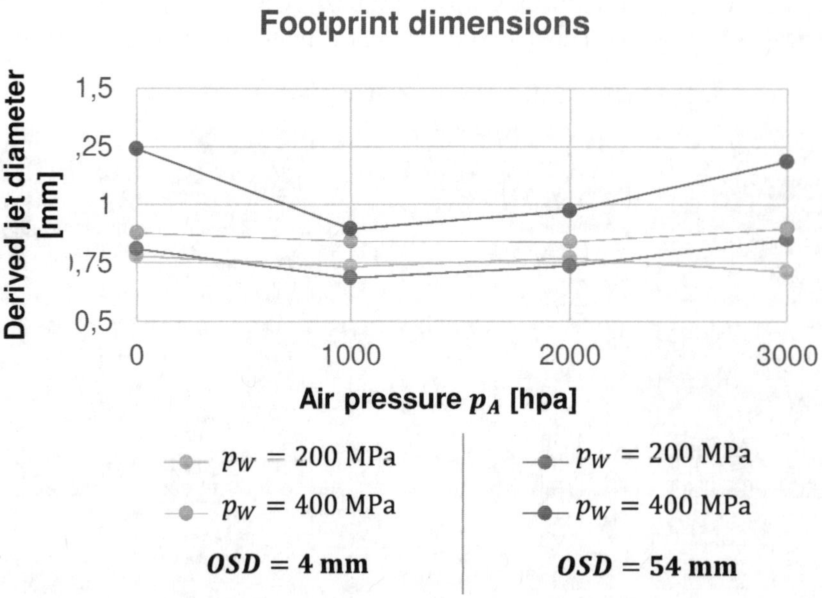

Figure 11. Overview of the footprint dimensions achieved under different levels of pressure conditions for both levels of SOD.

5. Summary and Outlook

In this study, a novel and simple modification of a waterjet tool was practically tested. Here, the fundamentals in jet geometry and erosion characteristics of an injection-type plain waterjet (iPWJ) were investigated for different modifications of the air inlet. From a blocked, up to a pressurized air supply into the iPWJ tool, both its effects on the free jet were visually compared and its footprints on an aluminum alloy were analyzed by SEM methods.

In the continuation of the work by [7] and [12], several conditions of the air inlet were analyzed. Besides the change in the free jet shape, the erosion could be altered. An increase in the erosion depth was seen at the low level of water pressure for both the atmospheric pressure p_A (1) = 1000 hPa and the mildly increased air pressure p_A (2) = 2000 hPa. The effective jet diameter was increased at the outer levels of air pressure, e.g., a blocked inlet p_A (0) or the high level of air pressure p_A (3) = 3000 hPa. While the widening of the free jet was visible at the high level of air pressure, it was not recognized at the blocked inlet. In contrast, the high level of water pressure p_W = 400 MPa revealed a local minimum of the footprint diameter at the air pressure p_A (1) and p_A (2), respectively. On the other hand, the same condition at the low level of water pressure p_W = 200 MPa created the most consistent erosion. It appears that a defined airflow into the iPWJ tool helps to focus the jet and vice versa. Overall, the shape of the free jet correlates well with the findings on the erosion results. The SEM images revealed a decrease in the size of the shallow erosion cavities by the high level of SOD, which was most notable for the outer levels of air pressure. This could be a sign that the jet was losing its coherence due to disintegration under atmospheric drag. However, at the low level of SOD, this could also be caused by changes in the recompression region. A summarization of the findings was given in the following:

- The shape of the free jet correlated well with the dimensions of the footprints.
- Jet stability varied based on the air input. A lowered stability was observed on the footprints at higher SOD.
- The effect of SOD on the size of erosion cavities was observed by SEM analysis of the machined specimen.
- A strong correlation between air pressure and jet diameter was observed for the low level of water pressure. The groove is deep and continuous for medium levels of air pressure. In contrast, both the blocked and highly pressurized conditions revealed isolated erosion pits with small erosion cavities and plastically deformed areas in between.
- The high level of water pressure resulted in pronounced and continuous grooves for all air pressure levels. A notable difference was observed in the apparent increase of groove width for the outer levels of the air pressure, namely a blocked inlet and a highly pressurized condition, respectively.

In the future, experiments should be done to optimize the phase composition of an iPWJ tool for distinctive applications. Here, the extended analysis of high-speed imaging or particle image velocimetry might be helpful. Further, the erosion capabilities of such adapted jet tools should be investigated more in detail to gain an understanding of the jet phase-material interaction. A preliminary evaluation of tool geometry, material removal rate, and surface characteristics on other materials might be necessary for modeling and integration of specific jet tools into computer-aided production. These developments could be promising for future applications of surface treatment with waterjets.

5. Acknowledgments

J. Poloprudsky would like to thank the Ministry of Education, Youth and Sports of the Czech Republic for supporting 3-month research stay at the Fraunhofer Institute for Production Technology IPT through Project No. CZ.02.2.69/0.0/0.0/18_053/0016933.

6. References

[1] Axinte, D.A., Karpuschewski, B., Kong, M.C., Beaucamp, A.T. *et al.*, 2014. High Energy Fluid Jet Machining (HEFJet-Mach): From scientific and technological advances to niche industrial applications, p. 751

[2] Foldyna, 2019. Recent Developments in Pulsating Water Jets. In: Hloch S., Klichová D., Krolczyk G., Chattopadhyaya S., Ruppenthalo-vá L. (eds) Advances in Manufacturing Engineering and Materials. Lecture Notes in Mechanical Engineering. Springer

[3] Jerman M., Zeleňák M. Lebar A. *et al.* (2021) Observation of cryogenically cooled ice particles inside the high-speed water jet. Journal of Materials Processing Technology v. 289

[4] Kong M.C., Axinte D., Voice W., 2011.Challenges in using waterjet machining of NiTi shape memory alloys: An analysis of controlled-depth milling. Journal of Materials Processing Technology

[5] Putz 2018 Putz M., Dix M., Morczinek F., Dittrich M., 2018.Suspension Technology for Abrasive Waterjet (AWJ) Cutting of Ceramics, p. 77

[6] T. Bergs, J.P. Borrmann, M. Schüler, T. Herrig, J.-E. Döring, Editor, 2019. Pure waterjet controlled-depth machining for stripping ceramic thermal barrier coatings on turbine blades. Elsevier B.V.

[7] Osman A.H., Mabrouki T., Théry B., Buisine D., 2004. Experimental analysis of high-speed air-water jet flow in an abrasive water jet mixing tube. Flow Measurement and Instrumentation 15:37–48

[8] Momber A.W, Kovacevic R. 1998. Principles of abrasive water jet machining. Springer.

[9] Momber A.W. 2001. Principles of abrasive water jet machining. Energy transfer during the mixing of air and solid particles into a high-speed waterjet: an impact-force study. Experimental Thermal and Fluid Science 25:31–41

[10] Huang C.Z., Hou R.G., Wang J., et al. 2007. Three-Dimensional Simulation of Liquid-Solid Two-Phase Flow Inside the Abrasive Water Jet Nozzle. KEM 329:329–334

[11] Himmelreich U. 1993. Fluiddynamische Modelluntersuchungen an Wasserabrasivstrahlen ((Fluid-dynamic model investigations on water-abrasive jets)), Ph.D. thesis Hannover: Fortschritt-Berichte VDI, Institut für Werkstoffkunde

[12] Bergs T, Schüler M, Dadgar M, et al. 2020. Investigation of Waterjet Phases on Material Removal Characteristics. Procedia CIRP 95:12–17.

[13] M. Schüler, T. Herrig, T. Bergs. 2022. A study on abrasive waterjet multi-stage machining of ceramics. Procedia CIRP 108:770-775.

7. Nomenclature

PWJ	Plain waterjet
AWJ	Abrasive waterjet
ASJ	Abrasive suspension waterjet
iPWJ	Injection-type plain waterjet
CFD	Computational fluid dynamic
v_{Jet}	Jet velocity
v_F	Feed rate
$\emptyset d_f$	Focus tube diameter
$\emptyset d_w$	Nozzle diameter
MPa	Megapascal as a unit of pressure
hPa	Hektopascal as a unit of pressure
SOD	Standoff distance
SEM	Scanning electron microscopy
EDS	Energy dispersion spectroscopy
SoD	Standoff distance
p_A	Air pressure
p_W	Water pressure

2022 WJTA Conference and Expo
November 2-3 • New Orleans, Louisiana

Paper

ANALYSIS OF WATERJET TREATMENT FOR HERBICIDE-FREE VEGETATION MANAGEMENT ON RAILWAY TRACKS

A. Schulte-Marxloh, U. Conrath
Department of Plant Physiology, RWTH Aachen University
Aachen, Germany

M. Schüler, T. Bergs
Fraunhofer Institute for Production Technology IPT
Aachen, Germany

C. Archut, N. Jendrny, C. Schindler
Institute for Rail Vehicles and Transport Systems, RWTH Aachen University
Aachen, Germany

ABSTRACT

Vegetation management is essential for the functionality and operational safety of railway tracks. Because the approval of the herbicide glyphosate on railway tracks will expire by the end of 2022 in Europe, there is an urgent need for alternative vegetation management strategies. An eco-compatible alternative could be the use of pressurized water, possibly supplemented with abrasives, to substitute current herbicides. This study examines two different approaches of waterjet methods to remove plants from railway tracks, either by cutting or defibering the aerial parts of plants. Here, a range of plant species and infrastructure components (e.g., cable insulation) were processed at varying process parameters. Furthermore, the benefit of using garnet abrasive for enhanced plant damage was examined. The experiments revealed that, with an appropriate parameterization of the waterjet process, plants could be effectively damaged without affecting the surrounding infrastructure.

Organized and Sponsored by WJTA®

1. INTRODUCTION

Vegetation management is essential for the functionality of railway systems and safety of rail traffic. According to Section 4 (3) of the German General Railway Act (AEG) [1], railway operators must construct and maintain railway infrastructure safely. In the last decades, most countries within the European Union (EU) primarily used chemical leaf and soil herbicides to keep the track area free of plants and ensure the functionality of the track. Glyphosate is currently the most commonly used leaf herbicide for effective vegetation control [2]. In the EU, the approval of the active ingredient glyphosate for vegetation control on railway tracks will expire by 2023 [3, 4], so operators of railway infrastructure are under political, legal, and social pressure to find glyphosate-free alternatives for vegetation control [2, 5]. Meanwhile, some focus on replacing glyphosate with a potentially more socially acceptable chemical herbicide such as pelargonic acid. Others are trying to avoid the use of herbicidal agents in general and therefore increasingly look for non-chemical methods [2, 5].

The suitability of multiple methods for future vegetation control is currently being investigated. This study is part of the project "Development of an Alternative Method for Vegetation Control on Railway Tracks" of the German Centre for Rail Traffic Research (DZSF) at the German Federal Railway Authority (EBA) and is funded by the Federal Ministry for Digital and Transport (BMDV). The aim of the project is the basic development or adaptation of non-synthetic chemical or alternative methods in vegetation management on railway tracks. A method or combination of methods shall be integrated into a prototype vehicle that can be used to test applicability. Initially, 18 alternative measures to glyphosate-based vegetation control were selected and evaluated theoretically for their suitability for vegetation control on railway tracks.

One of the promising methods for vegetation control is the treatment with pressurized water. The use of pressurized water is known for surface treatment in urban and industrial areas [6–8], is applied industry-specific for cutting materials [9–11], and is used in the track for rail head cleaning [12]. Based on the current potential analysis, the technology can be used promisingly in combination with other mechanical and electrophysical processes [13].

In general, chemical-free methods have their advantages and disadvantages. However, the success of the chosen method depends strongly on its implementation and the boundary conditions of the application. Vegetation control with waterjets allows application at a remote distance from the plant. That could be beneficial here. For example, in an area close to the rails, the plants still could be treated, while methods using a ground-level applicator, such as machine weeding or electrophysical treatment, must maintain a certain distance to the infrastructure components. In addition, pressurized water causes only minor involvements of natural and biodegradable substances, such as water or abrasive. Therefore, mechanical methods, which include the action of the water jet, are considered non-toxic and can be used in water protection areas. [5].

We carried out pilot experiments to determine appropriate parameters for reliable vegetation management with waterjets on railway tracks. In a collaboration with the Fraunhofer Institute for Production Technology IPT, the Department of Plant Physiology (iPP – RWTH Aachen University), and the Institute for Rail Vehicles and Transport Systems (ifs – RWTH Aachen University), common plants on railway tracks and railway components were treated in a high-pressure waterjet machine to determine crucial parameters such as pressure, application distance and nozzle position. The implementation of the method was done with and without abrasive in comparison. A suitable parameter range for the water jet process must be found to damage the plants without simultaneously affecting the integrity of the railroad infrastructure.

2. METHODOLOGY

To evaluate the effectiveness of the waterjet on the plant structure, a specified setup of typical plants in combination with infrastructure components was processed using an industrial waterjet machine. An H.G.Ridder Type Waricut HWE P2030 5-axis AWJ machine tool was used for experiments (**Figure 1**). Originally designed to cut complex structures, the machine system is capable of water pressure of up to 600 MPa. The machining range of the X-Y-Z axes is 2050 mm x 3050 mm x 400 mm, with a maximum linear feed rate of 20 m/min. This machine is equipped with two individual AWJ heads, each of them configured by different kinematics and z-axis. The machining unit used for the experiments allows the C axis to be rotated by +/- 540 degrees and the B axis to be tilted +/- 95 degrees. A UHDE Type 6045 intensifier pump was used, capable of water pressure of 600 MPa at a flow rate of 2.8 l/min, powered by a 45-kW electric motor. All components of this machine system were controlled by a SIEMENS Type Sinumerik 840D SL CNC control unit. Here, the waterjet tool was configured with a provided injection-type waterjet head, so that either a standard injection abrasive waterjet (AWJ) or an injection-type plain waterjet (iPWJ) can be used by entraining abrasive or disrupting the abrasive flow into the mixing chamber. For all configurations, a 0.25 mm water orifice and a 0.76 mm focus tube were used.

Figure 1. Waterjet machine and arrangement of the specimen.

The experiments were conducted based on design of experiments (DoE) methods. In the parameter studies, three water pressure levels p of 75 MPa, 150 MPa, and 300 MPa were investigated. It should be mentioned that the stand-off distance (SOD) of the AWJ nozzle over the specimen was adjusted to match the dedicated application. Here, the geometrical design of the track vehicle is limited to a distance of 300 mm above the ground [14]. Hence, the waterjet was applied at SODs of 100, 200, and 300 mm, respectively. This parameter pattern was carried out in two ways, once with and once without the addition of abrasive grit into the jet tool. Here, a garnet abrasive of mesh size #120 was used, entrained with a constant flowrate $\dot{m}_A$ = 200 g/min. The feed rate $v_F = 2000\ mm/min$ was set constant for this fundamental experiment, corresponding to a 1,2 km/h equivalent. Note, the last two factors were not varied further to make a fundamental decision regarding the subsequent system variant. Finally, the AWJ part was extended to an inclination angle of β = 45 deg in addition to the perpendicular β = 90 deg impact on the specimen surface. The design of the experiments primarily served to uncover requirements for further investigations, without neglecting the framework conditions of the subsequent application. **Table 1** gives an overview of the experiments, where the first number of the individual run represents the factor category that has been varied further to the three water pressure levels by the following letter.

Table 1. Parameters investigated

Run	Water pressure p [MPa]	Stand-off distance (SOD) [mm]	Abrasive mass flow $\dot{m}_A$ [g/min]	Inclination angle β [deg]
Run 1 a	75	300	-	90
Run 1 b	150	300	-	90
Run 1 c	300	300	-	90
Run 2 a	75	200	-	90
Run 2 b	150	200	-	90
Run 2 c	300	200	-	90
Run 3 a	75	100	-	90
Run 3 b	150	100	-	90
Run 3 c	300	100	-	90
Run 4 a	75	300	200 g/min	90
Run 4 b	150	300	200 g/min	90
Run 4 c	300	300	200 g/min	90
Run 5 a	75	200	200 g/min	90
Run 5 b	150	200	200 g/min	90
Run 5 c	300	200	200 g/min	90
Run 6 a	75	100	200 g/min	90
Run 6 b	150	100	200 g/min	90
Run 6 c	300	100	200 g/min	90
Run 7 a	75	300	200 g/min	45
Run 7 b	150	300	200 g/min	45
Run 7 c	300	300	200 g/min	45
Run 8	150	200	200 g/min	45
Run 9	75	200	200 g/min	45

A batch of individual workpiece materials represents a specimen that was processed for each parameter setup. First, the effect on the plant structure was determined with the aim of cutting or shredding parts of the individual plants with the water jet. Moreover, the effect of the primary or direct damage to the plant structure could lead to secondary damage by reduced biomass or height of the plants and possibly resulting in the death of the whole plant afterward. For this purpose, herbaceous (herb-Robert and ribwort plantain – stems 2-4 mm with leaves), grassy (upright brome – stems 2-3 mm), and partially woody plants (blackberry vines with an average diameter of 5 mm) were processed correspondingly. Second, typical infrastructure components from a railway track made of plastics were processed in analogy, because such materials are considered critical for waterjet application due to their softness. These were standard signal cables used in the track with a diameter of 20 mm and 32 mm, insulation of polyethylene, and a plastic cover made of 4 mm thick polypropylene [15].

For each run, the specimens were placed side by side on the test stand and processed in a single line for each parameter combination. All objects were removed, labeled, and analyzed visually. The evaluation includes measuring the cutting depth and the cutting width. In addition, the test plants were left to grow for one month to understand the post-effects of the treatment. A cause-effect analysis of the experimental section included the fatal losses of the complete plant or whether parts of the plant died or grew back.

3. RESULTS AND DISCUSSION

The study revealed that different results could be achieved on both the test plants and infrastructure components. Furthermore, the parameter range was limited where, on one hand, the plants were sufficiently damaged for vegetation control, and the infrastructure components were barely damaged on the other. The evaluation was conducted to find a fundamental waterjet configuration that allows effective damage to the plant structure while avoiding significant interference with the railway infrastructure components. It will serve as a basis for innovative approaches to vegetation management with water jets.

Results of iPWJ (no abrasive, $\beta = 90\ deg$)

When assessing the results achieved with the highest pressure level p = 300 MPa, it was found to have a devastating effect on plants. In several runs, blackberry vines were cut off or received up to 3-mm-deep cuts in their bark. In addition, the stems of the remaining test plants were cut off completely. However, even at the highest SOD = 300 mm and without abrasive added, significant damage to the infrastructure components was obvious (**Figure 2**). Regarding the depth of cut and fraying of surface, the damage increased with the smaller SOD = 200 and SOD = 100 mm respectively.

From these results, it was concluded that a pressure level of p = 300 MPa can be considered high for a corresponding application on the railway track. Due to the major damage, this factor level has been excluded from further experiments here. As an option, the highest pressure level could only be applied in track areas where cables, plastic covers, and components with similar hardness do not exist.

Figure 2. Results of p = 300 MPa and SOD = 300 mm (no abrasive, β = 90 deg).
Left: plastic cover, middle: signal cable, right: blackberry vine.

A medium pressure level p = 150 MPa damaged both blackberry vines at reduced distances (SOD = 200 and SOD = 100 mm, respectively) and the other test plants in a wide-ranging manner. However, the damage to the infrastructure increased at SOD = 200 mm (**Figure 3**) or below. Thus, for SOD = 300 mm or beyond, water pressure of 150 MPa seems to be adequate. At the same time, this distance should not fall below a minimum value of SOD = 200 mm in areas with infrastructure components.

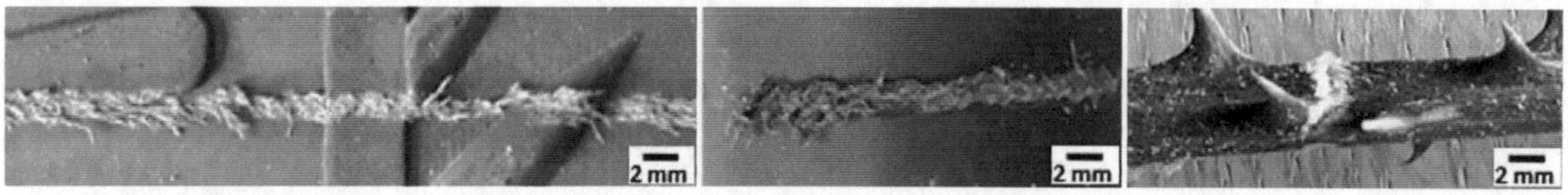

Figure 3. Results of p = 150 MPa at a SOD = 200 mm (no abrasive, β = 90 deg). Left: cover, middle: signal cable, right: blackberry vine.

At the lowest pressure level p = 75 MPa (**Figure 4**), almost no damage to the infrastructure was observed at SOD = 300 mm. With decreasing distance, the damage could be increased respectively. At SOD = 200 mm, the cable made of polyethylene was slightly damaged even at p = 75 MPa, but the cover made of the harder polypropylene was virtually not damaged. At the closest SOD = 100 mm, more severe damage occurred, eroding the insulation of the cable or fraying the plastic of the cover. Here, the effect on blackberry vines was barely sufficient, so the damage was unsatisfactory at SOD = 200 and 300 mm correspondingly. As an intermediate conclusion of the lowest water pressure, cutting the blackberry vines (5 mm diameter) was limited to the closest SOD = 100 mm.

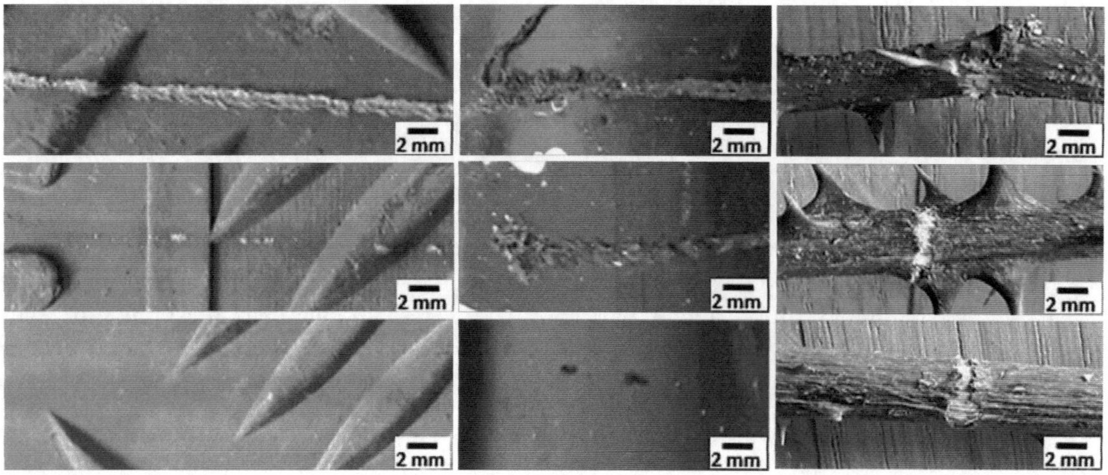

Figure 4. Results of p = 75 MPa at a SOD = 100 mm (top), 200 mm (middle) and 300 mm (bottom) (no abrasive, β = 90 deg). Left: cover, middle: signal cable, right: blackberry vine.

Summarizing the achieved results of iPWJ without using abrasive, the damage of even herbaceous plants was apparent at pressure levels between $p = 75$ and $p = 150$ MPa. Noticeably, the damage to the plants occurred here by defibering or cutting off the leaves and stems. In a comparison between the medium pressure level $p = 150$ MPa and the highest pressure level $p = 300$ MPa, the latter does not lead to remarkable advantages. A rise in pressure is usually accompanied by increased consumption of resources, higher maintenance costs, and more expensive equipment. Comparatively, the factor SOD showed great potential for optimization. On one hand, the highest SOD = 300 mm was, in some cases, on the limit of destroying the plants completely. On the other, a SOD = 100 mm was too close not to damage the infrastructure components. Here, a pressure level of $p = 150$ MPa can be seen as a maximum although the closest SOD = 100 mm initiated certain damage to the infrastructure components for all investigated pressure levels, e.g., by deep cuts or fraying of the insulation of the cables. Contradictory, for cutting woody plants such as blackberry vines, a small SOD = 100 mm may be necessary at the lowest pressure level. Herbaceous plants were processed sufficiently even at larger SODs. As an intermediate conclusion for iPWJ in vegetation control on railway tracks, the system should be capable of water pressure between $p = 75$ and 150 MPa.

Results of perpendicular AWJ ($\dot{m}_A$ = 200 g/min, $\beta = 90\ deg$)

For the AWJ experiments, the same DoE as for the aforementioned tests was considered, according to **Table 1**. Here, the previously used iPWJ configuration was connected to an abrasive supply, so that a standard AWJ configuration was accomplished. In analogy to the iPWJ tests, a comparison of the effect on both the test plants and the infrastructure components was made to evaluate the use of an abrasive.

At the highest pressure level $p = 300$ MPa, the occurred damage to the infrastructure components was different in comparison with iPWJ. Due to the interaction of the abrasive particles with the water jet, a larger area was interacted with. The effect changed from cutting or fraying to roughening or blasting of the surface, for an increased SOD. Especially at SOD = 100 mm, the damage was very severe, such that the material was heavily roughened and frayed (**Figure 5**). When processing blackberry vines, a similar roughening of the bark around the cut area was seen. The depth of cut ranged from 1 mm at SOD = 300 mm to 4 mm at SOD = 100 mm respectively. All other test plants were almost completely cut off or defibred at all SOD levels investigated. Similar to the industrial application of waterjets, the use of abrasive increased the cutting performance, in particular for the highest pressure level $p = 300$ MPa. Here, damage was already severe after a single pass, meaning that multiple treatments could harm the infrastructure components fatally. Correspondingly, the highest pressure level $p =$ 300 MPa could only be used in areas without sensitive infrastructure components or where such components will be protected.

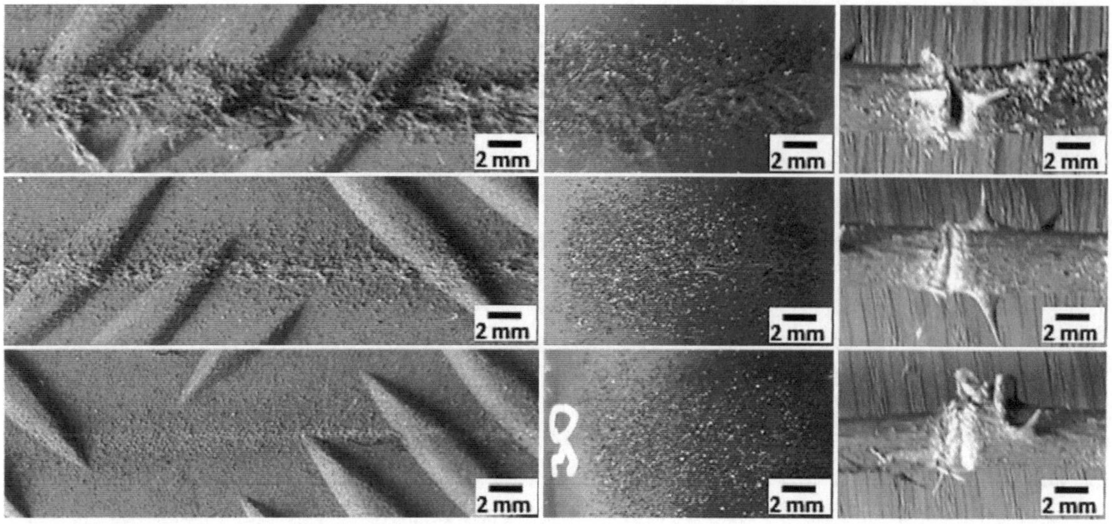

Figure 5. Results of AWJ (abrasive, β = 90 deg) at p = 300 MPa and SOD = 100 mm (top), SOD = 200 mm (middle) and SOD = 300 mm (bottom)
Left: cover, middle: signal cable, right: blackberry vine.

At a pressure level p = 150 MPa and SOD = 300 mm, it was shown that the damage to the infrastructure components decreased. Particularly with the polypropylene cover, barely any damage occurred on the surface. Only the surface of the cable showed roughening. However, the effect on blackberry vines also decreased. Thus, at p = 150 MPa and SOD = 300 mm, only the bark was damaged (**Figure 6**). As the SOD level was decreased, the depth of cut increased to 4 mm, but the damage to the infrastructure components also become more evident. Most test plants were either cut or defibred at all distances.

Figure 6. Results of AWJ (abrasive, β = 90 deg) at p = 150 MPa and SOD = 300 mm.
Left: cover, middle: signal cable, right: blackberry vine.

At a pressure level p = 75 MPa, little damage to the infrastructure was observed at all three distances, and no damage was caused at this pressure even at a SOD = 100 mm (**Figure 7**). But these settings could no longer damage the blackberry vines sufficiently, so the bark was only roughened. Also, most of the test plants were defibred at all distances, but not completely cut off.

In sum, the results of the treatments with abrasive revealed that for herbaceous plants, like the experiments without abrasive, a pressure p of 75 to 150 MPa effectively damaged the plants by crunching or cutting off. An increased pressure p = 300 MPa did not bring an advantage concerning the damage to plants but increased the resources needed.

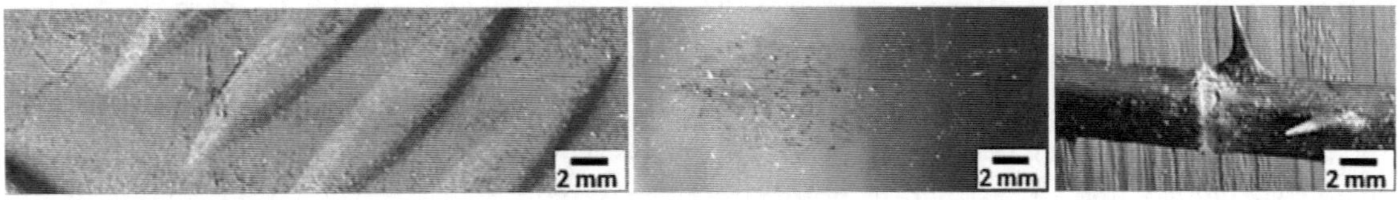

Figure 7. Results of AWJ (abrasive, β = 90 deg) at p = 75 MPa and SOD = 100 mm.
Left: cover, middle: signal cable, right: blackberry vine.

The SOD level appears to be an important aspect, even when entraining abrasive. Similar to iPWJ, the highest SOD = 300 mm turned out to be not sufficient for all plants. However, the SOD = 100 mm was too close for infrastructure components with a pressure level above p = 75 MPa, despite the better effect on the test plants. In comparison with the previous experiments achieved with iPWJ, the use of AWJ changed the results more to a roughening of the surface, even on the infrastructure components, but the leaves and stems of the plants also showed extensive damage coming from the abrasive interaction. Overall, it can be expected that more parts of the plants will dry out after treatment. When treating partially wooden plants such as mature blackberry vines, a narrow SOD = 100 mm would be necessary, but SOD can also be increased to cut or defibre herbaceous plants.

<u>Results of inclined AWJ ($\dot{m}_A$ = 200 g/min, $\beta = 45\ deg.$)</u>

In this last experiment, the nozzle was pitched at an inclination angle $\beta = 45\ deg$ at a constant SOD = 300 mm. In other words, the effective SOD was increased to achieve the influence of non-perpendicular jet impact within the geometrical limit of the envisaged test vehicle. The other parameters remain unchanged and abrasive was used. Focussing on the infrastructure components, **Figure 8** reveals that the damage could be decreased further in comparison with the foregoing results; only a slight roughening of the plastic cover and insulation was evident. Except at the highest pressure level p = 300 MPa, where the damage was more severe, both the medium and lowest pressure levels lead only to a minor impairment of the insulation material.

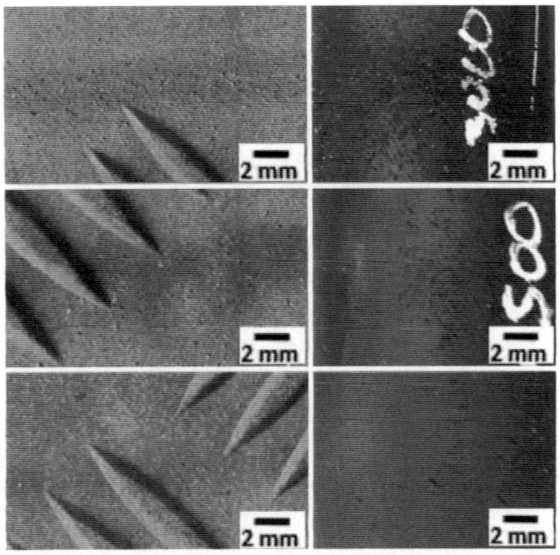

Figure 8. Run 7 (SOD = 300 mm, $\dot{m}_A$ = 200 g/min, $\beta = 45\ deg.$):
Bottom p = 75 MPa, middle p = 150 MPa, top p = 300 MPa. Left to right: cover and signal cable.

Figure 9 showed the effect of AWJ at a minimum pressure level p = 75 MPa, SOD = 200 mm and an inclination angle of the nozzle β = 45 deg. The effects shown here were on the test plants herb-Robert and ribwort plantain. Especially with these parameters, the effect of damage to the plants should be emphasized. The plant parts above the ballast were completely cut off, meaning the most severe damage to the plant. As can be seen in run 7 (**Figure 8**), the damage to the surface of the infrastructure components was reduced to a slight roughening of the surface. Thus, this combination of parameters showed the highest efficiency against damage to plants and at the same time only slight damage to infrastructure components. The inclination of the nozzle β = 45 deg obtained positive results because the damaging effect on plants increased due to the larger area covering the AWJ. The damage to the infrastructure components decreased because the impact is softened by the impact angle or the effective change in inclined SOD.

Figure 9. Run 8 and 9 (SOD = 200 mm, $\dot{m}_A$ = 200 g/min, $\boldsymbol{\beta} = \mathbf{45}\ deg$).
a+b: p = 150 MPa (a: herb-Robert, b: ribwort plantain),
c+d: p = 75 MPa (a: herb-Robert, b: ribwort plantain).

In analogy to the previous findings, the effective pressure range for vegetation control in railways tracks was estimated between a pressure level p = 75 MPa and p = 150 MPa for AWJ under β = 45 deg. Depending on the vegetation, the distance may be varied. For cutting thicker or partially woody plants a SOD of 100 mm is necessary, which could be increased in scenarios with little vegetation or smaller plants.

Revised parametrization range

The evaluation of the previous experiments was used to revise the waterjet parametrization range, as shown in **Table 2**. Combinations having unfavourable effects were crossed out. It also represents the evaluation of different combinations of parameters according to the damage to the infrastructure, meaning that the revision was focused on the factor levels of water pressure p.

Table 2. Overview of the revised parameters: Excluded pressure levels

Run	Water pressure p [MPa]	Stand-off distance (SOD) [mm]	Abrasive mass flow $\dot{m}_A$ [g/min]	Inclination angle β[deg]
Run 1	75 – 150 - ~~300~~	300	-	90
Run 2	75 – ~~150~~ - ~~300~~	200	-	90
Run 3	~~75~~ – ~~150~~ - ~~300~~	100	-	90
Run 4	75 – 150 - ~~300~~	300	200 g/min	90
Run 5	75 – ~~150~~ - ~~300~~	200	200 g/min	90
Run 6	75 – ~~150~~ - ~~300~~	100	200 g/min	90
Run 7	75 – 150 - ~~300~~	300	200 g/min	45
Run 8	150	200	200 g/min	45
Run 9	75	200	200 g/min	45

The highest pressure level p = 300 MPa was completely excluded. The results showed that the damage to the infrastructure components was too severe and that the effectiveness on the plants did not increase much compared to 150 MPa. At a pressure level p = 150 MPa and SOD = 200 mm or less, it will be necessary to check whether sensitive infrastructure components are located in the working area. In general, many of the test plants could be damaged by this pressure level. By entraining abrasive, the effect could also be intensified by roughening the plant surface. If thicker and partially woody plants are to be cut, the distance must be reduced accordingly. AWJ of a pressure level p = 75 MPa may be used throughout the track, according to achieved results, as the damage to the infrastructure components was very low. The effectiveness on plants was mainly achieved with smaller and thinner plants, thus an early treatment time during the vegetation period should be chosen. For thicker plants or denser vegetation, the SOD should be reduced further.

Finally, an inclination angle $\beta = 45$ deg of the nozzle was advantageous, because the tests showed that especially for dense vegetation the plants could be cut off quite well on the surface. Note that the effective length of the free-air jet was increased by $\sqrt{2}$ for inclined SOD, but the geometrical boundaries of the following prototype were considered. However, the cutting ability with thicker or partially woody plants decreased for $\beta = 45$ deg.

4. CONCLUSION

In this study, a fundamental investigation was conducted experimentally. The aim was to design and prototype mobile equipment for waterjet treatment that can be tested on railway tracks. In the experiments, standard injection-type waterjet equipment was used to obtain crucial factors in both system and process parametrization and to maximize the potential of waterjets for vegetation management. Thus, an optimal pressure level, the use of abrasives, and the stand-off distance were of special interest for the prototype, in addition to the power supply and weight of the necessary equipment.

The experimental results identified a specific parametrization range for injection-type waterjets at a pressure level between p = 75 MPa and p = 1500 MPa, in which the plant specimens were damaged sufficiently, without affecting the considered railway infrastructure. The minimum pressure level p = 75 MPa was promising when the SOD could be reduced. Beyond that, the pressure of 300 MPa caused severe damage to the infrastructure components, thus this high pressure level was excluded after evaluation.

In the future, this study will be extended toward an application on real-world railway lines. Here, it will be of special interest to consider annual (vegetational) seasons and the time frame when the plants could be treated best. To optimize the processing speed, an approach with multiple treatments could be promising. Further experiments will reveal how multiple treatments affect the condition of the infrastructure components and plants respectively. A longer observation of impacts and process optimization will be necessary to implement resource efficiency and sustainability.

5. ACKNOWLEDGMENTS

We acknowledge the project "Development of an Alternative Method for Vegetation Control on Railway Tracks" of the German Centre for Rail Traffic Research (DZSF) at the German Federal Railway Authority (EBA) is funded by the Federal Ministry for Digital and Transport (BMDV)

6. REFERENCES

[1] 1994. *Allgemeines Eisenbahngesetz vom 27. Dezember 1993 (BGBl. I S. 2378, 2396; 1994 I S. 2439), das zuletzt durch Artikel 10 des Gesetzes vom 10. September 2021 (BGBl. I S. 4147) geändert worden ist.*

[2] Nolte, R., Behrendt, S., Magro, M., Pietras-Couffignal, K., 2018. *HERBIE - Guidelines, State of the Art and Integrated Assessment of Weed Control and Management for Railways.* UIC-ETF, Paris.

[3] Bundesamt für Verbraucherschutz und Lebensmittelsicherheit. Liste der zugelassenen Pflanzenschutzmittel in Deutschland mit Informationen über beendete Zulassungen. https://www.bvl.bund.de/SharedDocs/Downloads/04_Pflanzenschutzmittel/psm_uebersichtsliste.pdf?__blob=publicationFile&v=15. Accessed 8 October 2021.

[4] The European Commission, 2017. *Commission Implementing Regulation (EU) 2017/2324 of 12 December 2017 renewing the approval of the active substance glyphosate in accordance with Regulation (EC) No 1107/2009 of the European Parliament and of the Council concerning the placing of plant protection products on the market, and amending the Annex to Commission Implementing Regulation (EU) No 540/2011.*

[5] Pietras-Couffignal, K., Below, M., Yilmazer, P., Nolte, R., Schuh, T., 2021. *Future vegetation control of European Railways State-of-the-art report (TRISTRAM Final Report)*, Paris.

[6] A. Chillman, M. Hashish, M. Ramulu, C. Lavender *et al.*, 2011. ENERGY BASED EVALUATION OF WATERJET PEENING FOR INDUSTRIAL APPLICATION.

[7] Bodnarova, L., Sitek, L., Foldyna, J., Jarolim, T. *et al.*, 2018. APPLICATION OF WATER JET TECHNOLOGY FOR CONCRETE REPAIR *2018, p.* 2385.

[8] C. K. Van Name, J. R. Wilburn, 2015. Mechanical Tube Cleaning: The Emerging Trend in Maintaining Heat Exchanger Efficiency.

[9] Bergs, T., Schüler, M., Arntz, K., Herrig, T., 2019. Technological Analysis of Different Abrasive Water Jet Machining Strategies for Blisk Roughing.

[10] Axinte, D.A., Karpuschewski, B., Kong, M.C., Beaucamp, A.T. *et al.*, 2014. High Energy Fluid Jet Machining (HEFJet-Mach): From scientific and technological advances to niche industrial applications.

[11] Henning, A., Miles, P., Stang, D., Editors, 2011. *Efficient Operation of Abrasive Waterjet Cutting in Industrial Applications.*

[12] Network Rail. Seasonal track treatment and weather support fleet - Network Rail. https://www.networkrail.co.uk/running-the-railway/looking-after-the-railway/our-fleet-machines-and-vehicles/seasonal-track-treatment-and-weather-support-fleet/. Accessed 24 September 2021.

[13] Archut, C., Jendrny, N., Schulte-Marxloh, A., Eberius, M., Conrath, U., Schindler, C., 2022. *Entwicklung eines Alternativverfahrens zur chemischen Vegetationskontrolle auf Gleisanlagen*, Veröffentlichung in Vorbereitung.

[14] 1967. *Eisenbahn-Bau- und Betriebsordnung (EBO).*

[15] Lichtberger, B., 2010. *Handbuch Gleis: Unterbau - Oberbau - Instandhaltung - Wirtschaftlichkeit,* 3rd edn. PMC Media House, 2010.

7. NOMENCLATURE

AWJ	Abrasive Water Jet
iPWJ	injection-type Plain Water Jet
EU	European Union
β	Inclination angle
$\dot{m}_A$	Abrasive mass flow
p	Water pressure
SOD	Stand-off distance
deg	degree as unit of angle
kW	Kilowatt
MPa	Megapascal
mm	Millimeter

2022 WJTA Conference and Expo
November 2-3 • New Orleans, Louisiana

Paper

BENCHMARK OF ABRASIVES FOR DIFFERENT APPLICATIONS

M. Schüler, R. Day, T. Bergs
Fraunhofer Institute for Production Technology IPT,
Aachen, Germany

ABSTRACT

The industrial use of abrasive waterjet (AWJ) technology has its strength in the machining of conventionally difficult-to-machine materials where the use of abrasives is obligatory. Since the use of AWJ technology, garnet abrasive has become the standard due to its performance and availability. Facing the impending shortage of natural raw materials, rising prices, and increasing environmental requirements, users of waterjet technology must optimize their processes. A wide range of individual machining tasks and materials to be processed holds great potential for tailoring the consumption of valuable resources and costs in the future.

A fundamental comparison of different abrasive grit is made to contribute to the efficient use of solids for dedicated applications and to identify resource-efficient alternatives. For this purpose, an overview is given of the abrasive performance in AWJ machining. The experiments will be examined for both AWJ cutting through (CT) and controlled-depth machining (CDM) or AWJ milling techniques on two heat-treated modifications of a 42CrMo4 steel alloy. Finally, a fundamental discussion of grit properties on the systematic change of size distribution serves as a basis to meet the requirements of reusability and their attractiveness for future waterjet production.

Organized and Sponsored by WJTA®

1. INTRODUCTION

The effective processing of high-performance materials is vital in all knowledge-based industrial sectors. Similar to all machining approaches, choosing the right tool is fundamental for competitiveness and decides between profit or losses [1]. On par with the resource-effective usage of the tool, resource savings and recycling processes are already upcoming [2]. In addition to the optimization of existing processes, the use of non-conventional alternatives promises high potential in optimization of production [3].

One of the most flexible, non-conventional production techniques for cutting these materials is abrasive waterjet (AWJ) cutting [4]. Cutting through the entire workpiece (CT) has been the standard application of waterjet in the industry since its introduction [5]. In addition, the potential of controlled depth machining (CDM) [6], also known as abrasive waterjet milling, was demonstrated decades ago [7]. Both machining strategies are pioneering for special applications of AWJ technology. Several aspects of tool-material interaction must be considered. For AWJ cutting high-strength materials, the addition of abrasive materials is mandatory [8]. By adopting the abrasives to the material, the waterjet can be used extremely flexible in a variety of machining tasks [9, 10].

With the further spread of water jet technology and the overall growing demand from other industries, the natural resources of garnet abrasives are becoming increasingly scarce. Garnet production has quadrupled in the last 15 years to 1.2 million metric tons per year worth $340 million dollars [11]. The raw material comes from either alluvial (beach) or mining (rock) resources from areas mainly in Australia, South Africa, the United States, China, and India [12]. Since garnet production and export have been increasingly regulated in recent years [13], rising market prices and thus higher costs in production are ubiquitous. In addition, the processing of special materials also produces waste with environmentally critical pollutants - heavy metals and phenols - which were previously discharged directly into the sewage system. Consequently, the cost-efficient recovery and treatment of the waste materials is an obvious and yet necessary challenge to secure the competitiveness of AWJ technology.

Garnets has been used almost exclusively on a one-time basis, although a large proportion can be reused from a technological point of view [13, 13–15]. It is expected that garnet can be partially substituted by alternative materials or recycling. Approximately 50 % of the cast-off garnet grit can be reused [15–17]. Alternative abrasives are promising for special applications [18]. Alternatives can either be of natural origin, e.g. other stone materials, or synthetically produced by melting involved materials, e.g. white fused alumina. However, the involved maintenance costs due high wear on system parts must be considered and ccontamination on the machined surfaces should be well-thought-out [19, 20]. Besides the cost aspect, the origin of raw materials impacts the ecological footprint. Of certain interest are alternatives made of other waste products from industries, e.g. slags from metal production. Nevertheless, the possibility of reuse before final disposal is one possibility to save resources.

2. EXPERIMENTAL SETUP

This study compares different types of abrasives and their erosion performance on a steel alloy. Two categories of abrasives were investigated. First, an assortment of garnet abrasives as a standard for AWJ cutting. Second, a choice of alternatives as a series of natural or synthetic minerals. Natural sources of abrasive grit include garnets from diverse regions of the world. Differences in grain shape and further composition also depend on the extraction method, e.g. whether it originates from mining or alluvial (beach). Alternative abrasives were chosen from different slags as waste products from other industrial branches and natural minerals. All abrasives were selected within a particle size distribution according to #80 mesh. A detailed overview of the average material properties of the garnet abrasives and the alternative abrasives is given in Table 1 and Table 2, respectively. It should be noted that all abrasives have a different bulk density, and the size of individual particles is subject to natural variations. Hence, control sieving was conducted before usage. The individual mass flow was calibrated to $\dot{m}_A = 100$ g/min for all abrasives by check weighing. Here, an H.G.Ridder Type Waricut HWE P2030 5-axis AWJ machine tool in combination with a UHDE Type 6045 intensifier pump was used for experiments. All components of this machine system were controlled by a SIEMENS Type Sinumerik 840D SL CNC control unit. A summary of the machine configuration and individual parametrization is given in Table 3.

A chromium-molybdenum 42CrMo4 steel alloy (AISI 4140) as listed in Table 4 has been chosen as the workpiece material because this steel offers a wide range of either ductile or brittle-hard conditions by heat treatment. This is obtained with the same base material and thus without significant change in material density. Hence, this study does not include other materials. The experiments were carried out by machining two different structural modifications:

- 42CrMo4 +A: Pearlitic structure (annealed, ~ 30 HRC)
- 42CrMo4 +QT: Martensitic structure (quenched, tempered, ~ 55 HRC)

To contribute to future applications of waterjet machining, a comparison is made between waterjet cutting through (CT) the entire workpiece and emerging approaches to controlled depth machining (CDM) techniques as applied to AWJ milling. Both strategies were applied to the specimen as shown in Figure 1. For CT, the cutting performance is evaluated as the cutting ability of a wedge-shaped workpiece with increasing cutting thickness until continuous cutting is no longer possible. The CT cutting performance is then used to discuss the kerf footprints achieved with CDM. Here, the material removal rate (MRR), the maximum height S_z of the surface structure in the kerf bottom as a quality criterion, and the average values of the footprint profile by its width w_P and depth t_P are evaluated to describe the engaged AWJ tool. Visual and optical methods are predominantly used for evaluation. While the CT cut lengths were determined visually, an optical micro coordinate measuring system type Alicona InfiniteFocus G5 was used for the analysis of the CDM footprints following the EN ISO 25178 standard.

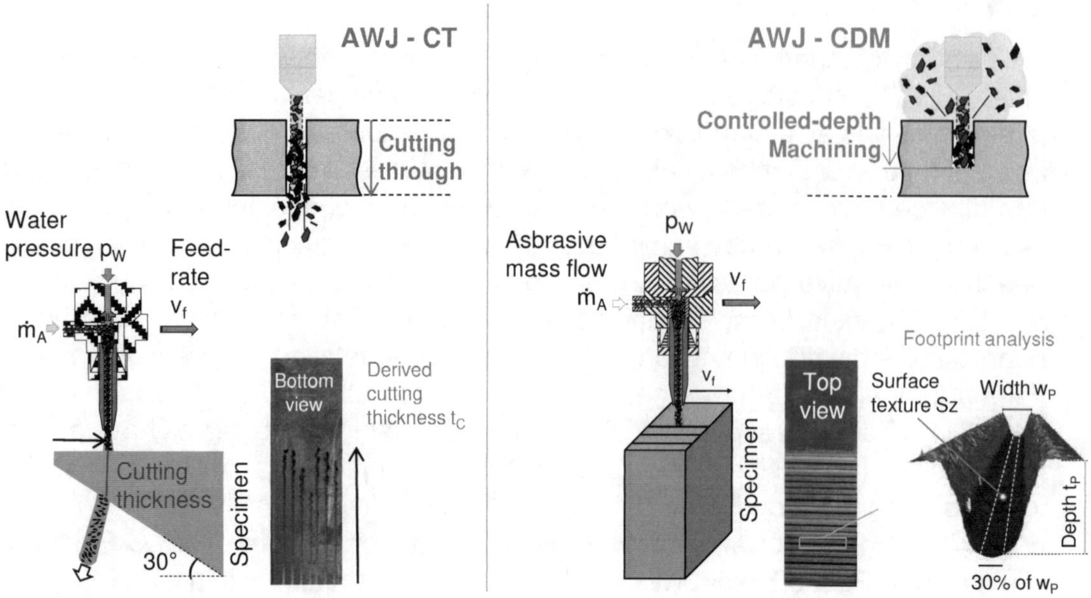

Figure 1: AWJ Machining strategy and experimental procedure. An evaluation was made for cutting through (CT) the entire specimen and analyzation of the AWJ footprint produced by controlled-depth machining (CDM) technique.

Table 1: Machine configuration and parametrization for CT and CDM experiments.

Machine configuration	***CT Parametrization***	***CDM Parametrization***
$\emptyset d_O = 0.250\,mm$	$p_W = 400\,MPa$	$p_W = 400\,MPa$
$\emptyset d_F = 0.76\,mm$	$\dot{m}_A = 100\,g/min$	$\dot{m}_A = 100\,g/min$
$SoD = 4\,mm$	$v_F = 25\,mm/min$	$v_F = 1250\,mm/min$

Table 2: Physical properties and composition of garnet abrasives.

	Beach SA	***Beach AU***	***Rock US***	***Rock CN***
Hardness	7.5-8.0 mohs	7.5-8.0 mohs	7.5-8.5 mohs	7.5-8 mohs
Grain shape	sub angular	sub angular	angular	angular
Specific weight	4.1 g/cm³	4.1 g/cm³	4.0 g/cm³	4.0 g/cm³
Composition	95 % Almandine 1 % Pyroxene 5 % Ilmenite	93 % Almandine 5 % Pyroxene 1 % Ilmenite	94 % Almandine, Pyrope, Grossular 6 % other minerals	90% Almandine 5% Pyroxene 2% Ilmenite

Table 3: Physical properties and composition of alternative abrasives.

	Coal Slag	Copper Slag	Steel Slag	Granite	Grey-wacke	Olivine
Hardness	7 mohs	6.5 mohs	7.5 mohs	7 mohs	7 mohs	7 mohs
Grain shape	angular	iso-metric	angular	sub angular	sub angular	sub angular
Specific weight	2.5 g/cm^3	3.6 g/cm^3	3.6 g/cm^3	3.3 g/cm^3	3.3 g/cm^3	3.3 g/cm^3
Composition	50 % SiO_2 28 % Al_2O_3 9 % Fe_2O_3 6.5 % CaO 2.5 % CaO 2.5 % CaO	53 % FeO 6 % Al_2O_3 5 % CaO 1 % K_2O 2 % MgO 1 % S	32 % SiO_2 29 % Al_2O_3 20 % MgO 10 % Cr_2O_3 5 % Fe_2O_3 4 % CaO	50 % Orthoclase, potassium feldspar 20 % Plagioclase feldspar 20 % Quartz 10 % Biotite	40 % Quartz 35 % Feldspar 15 % Mica 10 % Chlorite, carbonates	48 % MgO 40 % SiO_2 11 % FeO 1 % Al_2O_3

3. EVALUATION OF RESULTS

A comparison of the cutting performance for abrasive waterjet (AWJ) was made for each abrasive grit on two heat-treated modifications of an 42CrMo4 specimen by either AWJ cutting-through (CT) the entire workpiece or by an analysis of the footprints created by controlled-depth machining (CDM), like waterjet milling. In the pursuing evaluation of the experiments, several aspects of the involved material interaction must be considered. Both the kinetic and physical properties of the abrasive particles interacting with the workpiece material were known to be of fundamental importance for erosion. Therefore, a basic distinction should be made between the stress on the abrasives in CT, mainly due to sliding wear, and CDM correspondingly due to impact wear. Since this approach serves as a reference for further investigations on abrasive properties in AWJ machining, trademarks or brand names have been omitted. Instead, the type and origin of the abrasive was indicated.

In the first section of the CT assessment, a reference value was first determined within the garnet abrasives. Here, the percentage deviations of the most performant garnet abrasive were used. An average value was then taken from the garnets, which was then used as the new reference value for evaluating the alternative abrasives. The respective reference values were formed within a material modification and only used within it. In the second section of the CDM assessment, a similar procedure was followed. To benchmark alternative abrasives, a direct comparison was made against the garnet abrasives as being the standard in waterjet technology.

3.1 CT performance

In the first section of the evaluation, the cutting performance of common garnet abrasives was considered and compared with alternative abrasive materials. Other evaluation criteria, such as the characteristics of jet lag, the formation of burrs at the exit, or the edge quality, were not part of this study. Instead, a reference of enrolled garnet abrasives was selected by the most performing garnet within a modification of the 42CrMo4 steel.

The annealed 42CrMo4 +A (Figure 2 left) reveals a very similar cut-off capability for garnet abrasives. The US-origin rock garnet (Rock US) was the best performing abrasive, corresponding to a cutting thickness t_C = 36.1 mm. It serves as a reference in the following comparison, closely followed by the Australian beach garnet (Beach AU) with about 98.1 % ability of its predecessor. This was immediately followed by South African beach garnet (Beach SA) with about 94.7 % cutting performance in comparison and Chinese rock garnet (Rock CN) of 91.7 % respectively. Overall, the differences within the 42CrMo4 +A were in the range of approx. 8 % among each other and most likely within the range of natural fluctuations. On the other hand, deviations from the previous result were obvious at the quenched and tempered 42CrMo4 +QT (Figure 2 right). Although the previously rated high-performance garnets were also in the upper range of the cutting performance here, the type of material extraction was conspicuous. Both representatives of the mined rock garnet were above those from beach exploitation in their performance. Compared to the best-performing Rock CN with a cutting thickness t_C = 29.0 mm (reference) and Rock US (98 %), the Beach AU (96.5 %) was slightly ahead of the Beach SA (91.4 %). Despite the slightly different results and the overall reduced cutting performance at 42CrMo4 +QT, the range of deviations among each other was in a similar order of magnitude as for the previous 42CrMo4 +A.

The evaluation of alternative abrasives considered the average value within the preceding material modification of garnets as reference. This allows the comparison of the respective abrasive with the typical result of a garnet. Alternative abrasives were grouped into synthetically produced fused slags and natural minerals or stones. Melted Coal Slag is a synthetic mineral resulting from energy production by coal combustion. Copper Slag or copper silicate slag is produced from the corresponding melting process. Similarly, Steel Slag matches to a waste product in iron production. On the side of natural minerals, local rock types were also considered. Granite has hardly been used as an abrasive so far and is mainly used in the construction sector, as is Greywacke. Olivine, on the other hand, is a naturally occurring metal silicate from mining areas in India and China, which is also used industrially as an abrasive.

When cutting 42CrMo4 +A with the alternatives (Figure 3), the group of slags was fundamentally below the average performance of the garnets ($\overline{t_C}$ = 30.3 mm), but still above the natural minerals. Coal Slag was in first place with an equivalent to the garnet reference of 85.7 %, immediately followed by Steel Slag and Copper Slag (79.1 % and 73.9 %, respectively). For the natural minerals, Olivine still achieves 58.8 %, with Granite and Greywacke again halving the cutting performance further to around a third to that of garnets (31.3 % and 29.7 %, respectively). On the other hand, at the

42CrMo4 +QT, the group of slags performs similarly, although in some cases the differences become more obvious. Here, the reference value for comparison was t_C = 20.0 mm. The Steel Slag was in first place with 83.8 %, followed by Copper Slag and at a greater distance Coal Slag (66.7 % and 46.3 %, respectively). However, Olivine performs slightly better than on the annealed counterpart (62.0 %), as does Granite (32.4 %). Greywacke achieved a further drop with only 25 % compared to the garnet reference.

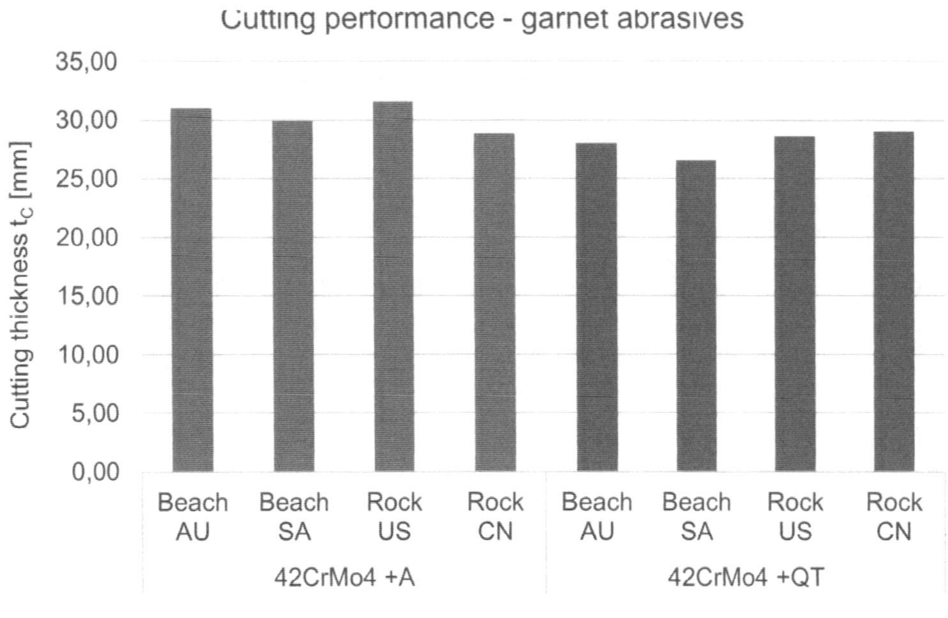

Figure 2: Cutting through (CT) performance of garnet abrasives. Derived cutting thickness t_C on a heat-treated 42CrMo4 steel. The results on the annealed 42CrMo4 +A are on the left side, the quenched and tempered 42CrMo4 +QT is on the right.

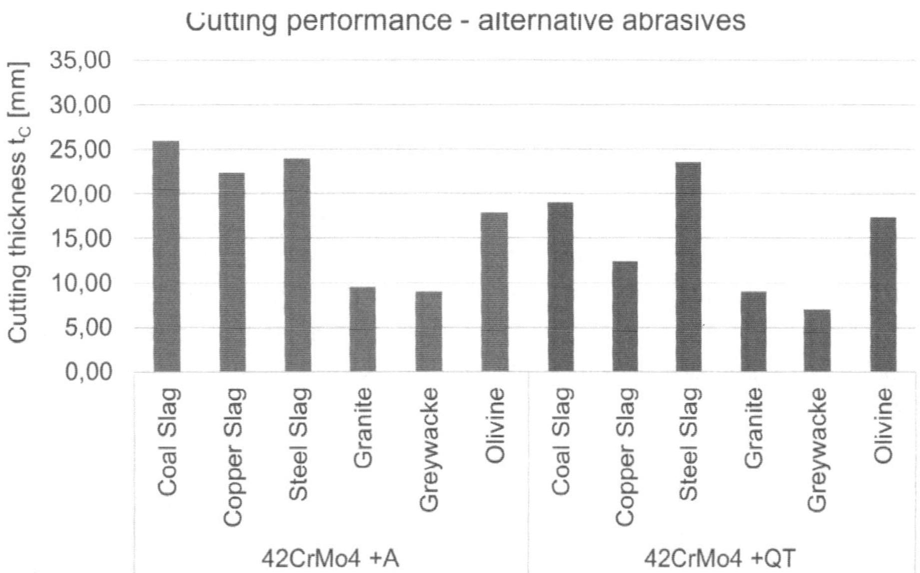

Figure 3: Cutting through (CT) performance of alternative abrasives. Derived cutting thickness t_C on a heat-treated 42CrMo4 steel. The results on the annealed 42CrMo4 +A are on the left side, the quenched and tempered 42CrMo4 +QT is on the right.

As an intermediate conclusion for the CT experiments, the garnet grits performed very similar at the respective material. The results only slightly differ according to the composition and physical properties of the abrasives. On the 42CrMo4, the mined rock garnets, typically characterized by sharp grains, performed above their alluvial counterparts. An initial difference could be made to their origin and exploitation method. However, the results quantified possible alternatives for substitutes. Here, the results differed more depending on the material modification of the specimen in contrast to the physical properties of the abrasives. The slags were distinguished by sharp grains, but with lower hardness compared with their counterparts. In the scope of AWJ CT, the garnet grit represents a reference against which alternative abrasives must measure themselves.

3.2 CDM characteristics

In analogy to the CT experiments, the results achieved with the garnet abrasives were compared categorically for CDM. Once, the category of garnet abrasives was first compared with each other. Then the alternative abrasives were benchmarked with the average of the garnets within a modification of the 42CrMo4 as a reference. In addition to the determination of the cutting performance based on the material removal rate (MRR), the machining quality was characterized here based on the maximum height S_z of the derived profile at the bottom of the footprint. The lower this value, the more uniform the achieved erosion, which corresponds to better surface texture. Basic information on the engaged AWJ tool geometry was finally derived by depth t_P and width w_P of the footprint.

The MRR on the ferritic-pearlitic microstructure of the annealed 42CrMo4 +A by CDM revealed a comparable ranking of the garnet abrasives as on the CT counterpart (Figure 4). In terms of performance, the rock garnet from the US (Rock US) was again in first place with an average MRR of around 545 mm^3/min and served as the reference on this material modification. The second highest result of 85 % was already about a quarter below this value and was achieved by the Australian beach garnet (Beach AU). The other two garnet abrasives were grouped close together in terms of removal rate. The Chinese rock garnet (Rock CN) achieved around 71.3 % of the reference value, while the South African beach garnet (Beach SA) achieved 62.6 %. On the martensitic structure of the quenched and tempered counterpart 42CrMo4 +QT, however, the mined Rock CN set the reference utilizing an average MRR of around 465 mm^3/min against the result of the also mined Rock US. The alluvial Beach AU and Beach SA were again quite a bit below in terms of machining performance, with 73.7 % and 69 %, respectively.

Deviating from the previous results, the characterization of the surface texture achieved with garnet grit by the maximum height S_z behaved detached from the CT performance (Figure 5). On the 42CrMo4 +A, a very uniform erosion due to the lowest S_z = 359 µm was achieved with Beach AU. This corresponds to only 56.5 % of the corresponding reference from the MRR results set by Rock US. Analogously, a correlation between the grain structure of alluvial and mined garnet was found. Beach SA (73.3 %) and Rock CN (154 %) varied further to reference. However, on the

42CrMo4 +QT the overall differences of S_z were significantly higher. Here, the reference value of Rock CN forms a minimum (S_z = 218 µm). Contrary, the result of the mined counterpart Rock US revealed a maximum of 271 %. The two remaining Beach AU and Beach SA were close together in the midfield with 205.6 % and 208.2 %, respectively.

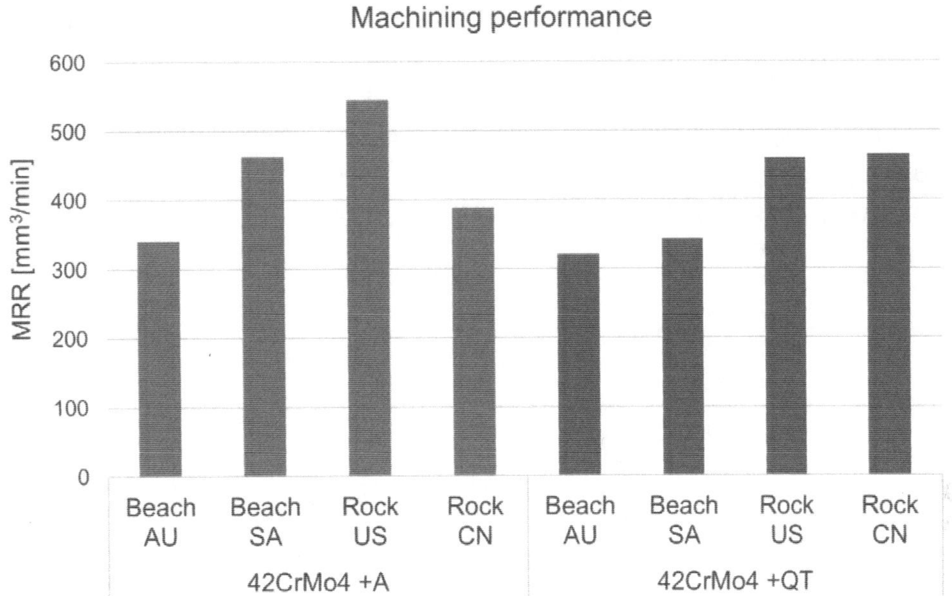

Figure 4: Controlled-depth machining (CDM) with garnet abrasives. Machining performance by material removal rate (MRR) on a heat-treated 42CrMo4 steel. The results on the annealed 42CrMo4 +A are on the left side, the quenched and tempered 42CrMo4 +QT is on the right.

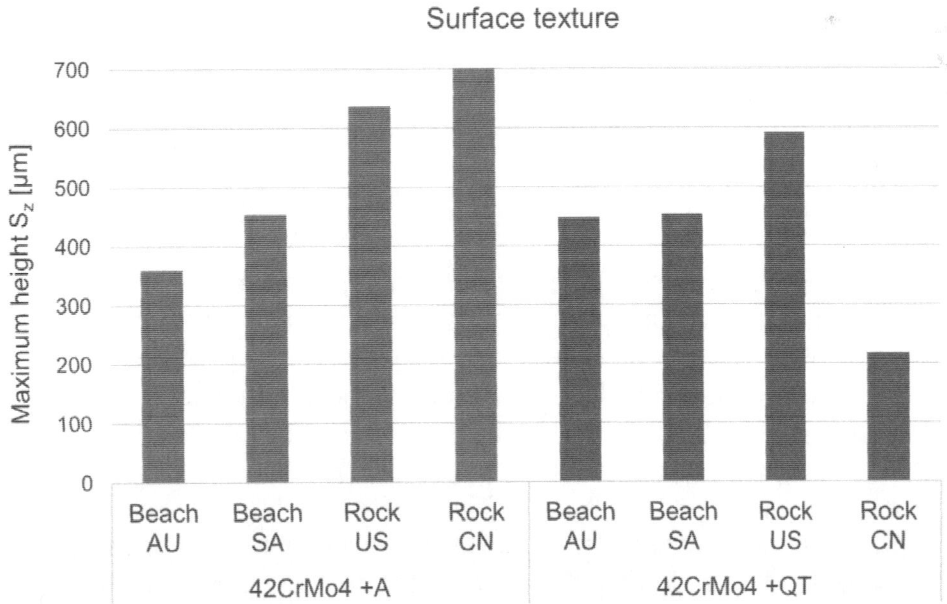

Figure 5: Controlled-depth machining (CDM) with garnet abrasives. Surface texture by the maximum height S_z on a heat-treated 42CrMo4 steel. The results on the annealed 42CrMo4 +A are on the left side, the quenched and tempered 42CrMo4 +QT is on the right.

The geometric dimensions of the footprints (Figure 6) by the mean depth t_P achieved with garnets on 42CrMo4 +A were predominantly in line with the associated courses of MRR. For both Beach AU (65.3 %) and Rock CN (62.2 %), the deviations to the reference Rock US (t_P = 732 µm) were practically identical. Beach SA with a ratio of 116.5 % revealed a slightly more pronounced depth in comparison. However, the corresponding results on the 42CrMo4 + QT varied more. The reference Rock CN (t_P = 440.0) previously set in the CT evaluation was approximately on par with Beach AU (102.8 %) on this hardened material modification. In contrast, the Rock US (181.8 %) and the Beach SA (194.0 %) achieved significantly greater depth values. Although the mean widths w_P on the 42CrMo4 +A showed only slight fluctuations within 3.6 % above the reference value, however, a minor reduction to smaller values within 3 % was found on the 42CrMo4 +QT.

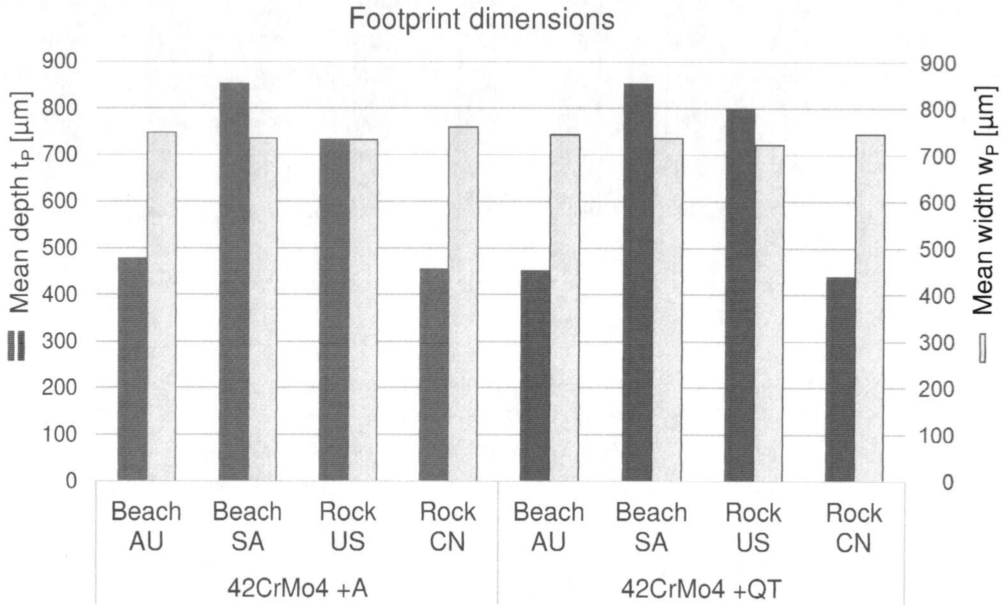

Figure 6: Controlled-depth machining (CDM) with garnet abrasives. Footprint dimensions by the mean depth t_P (left column) and the mean width w_P (right column) on a heat-treated 42CrMo4 steel. The results on the annealed 42CrMo4 +A are on the left side, the quenched and tempered 42CrMo4 +QT is on the right.

A comparison of the performance of alternative abrasives by MRR was similarly done with the average value of garnet grit on the respective material modification as a reference (Figure 7). Strong differences to the previous CT results were evident here. On the 42CrMo4 +A machined with the synthetic minerals, both the Coal Slag (63.3 %) and the Steel Slag (62.1 %) showed a similar performance equivalent to the garnet reference ($\overline{MRR}$ = 434.4 mm³/min). The Copper Slag (22.4 %) was inferior and almost on par with Granite (21.4 %) and Olivine (20.4 %). Greywacke (11.4 %) showed a strikingly low performance here. However, a significant performance decrease was also observed on the martensitic 42CrMo4 +QT. Both the Coal Slag (7.2 %) and the Copper Slag (2.1%) showed a strong decline in MRR in comparison to the associated garnet reference ($\overline{MRR}$ = 397.3 mm³/min). However, the Steel Slag still performed similarly well (62.2 %) here, so did Granite (19.7 %). The remaining Greywacke

(6.4 %) and Olivine (7.4 %), also suffer a significant decrease in removal rate. It gave the impression that some results between the materials behaved in opposite ways.

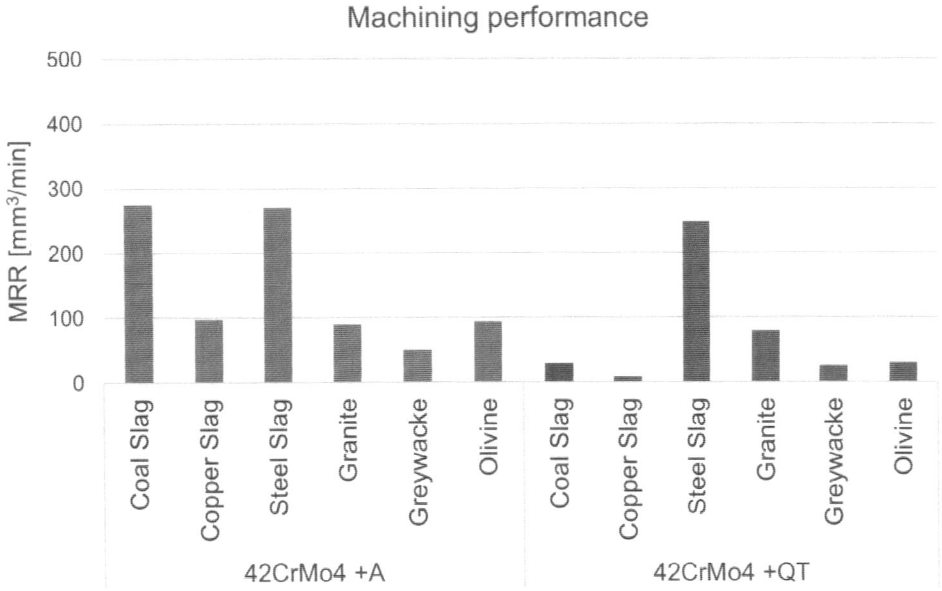

Figure 7: Controlled-depth machining (CDM) with alternative abrasives. Machining performance by the material removal rate (MRR) on a heat-treated 42CrMo4 steel. The results on the annealed 42CrMo4 +A are on the left side, the quenched and tempered 42CrMo4 +QT is on the right.

In the evaluation of surface texture, all alternative abrasives showed a significant reduction of the maximum height S_Z (Figure 8). On the 42CrMo4 +A, the values of S_Z in the group of slags were very close. Here, the equivalent to the garnet reverence ($\overline{S_Z} = 608.3\ \mu m$) was 37.1 % for Coal Slag, 35.8 % for Copper Slag and 39.5 % for Steel Slag. Slightly above, but still below the garnet grit, were Olivine with 45.9 %, Granite with 65.3 %, and finally Greywacke with 77 %. On the 42CrMo4 +QT, one exception was achieved with Steel Slag (14.6 %) against the reference value of $\overline{S_Z} = 429.5\ \mu m$. The remaining minerals minimized the results of S_Z on this material further: Coal Slag (1.7 %) and Copper Slag (0.5 %) showed very good results, and Greywacke (1.5 %) and Olivine (1.7 %) had comparable expressions, followed by Granite (4.6 %).

The footprint dimensions for alternative abrasives were strictly following the corresponding MRR results (Figure 9). On the 42CrMo4 +A, the average depth t_P achieved with the slags in comparison with the garnet reference ($\overline{t_P} = 630.3\ \mu m$) was 26.5 % for Copper Slag, 71.7 % for Steel Slag and 75.1 % for Coal Slag, respectively. A lower depth t_P was achieved when using natural minerals, equivalent to 12.5 % for Greywacke, 21.9 % for Granite and 28 % for Olivine. The corresponding results on the 42CrMo4 +QT were even closer. Here, the reference was $\overline{t_P} = 636.5\ \mu m$. The group of slags corresponded to 2.7 % for Copper Slag, 6.9 % for Coal Slag and 64.1 % for Steel Slag. A further deviation, but with overall lower values, were achieved for Greywacke (6.8 %), Granite (19.4 %) and Olivine (8.9 %). However, the corresponding values of the mean kerf width w_P were practically on par with those of the garnet grit.

On the 42CrMo4 +A, the deviations against the garnet reference ($\overline{w_P} = 744.0\ \mu m$) were within +/- 2 %. For 42CrMo4 +QT, the majority was within +/-3 %. Here, an exception was found for Copper Slag, which was 9% smaller than the garnet reference $\overline{w_P} = 736.2\ \mu m$.

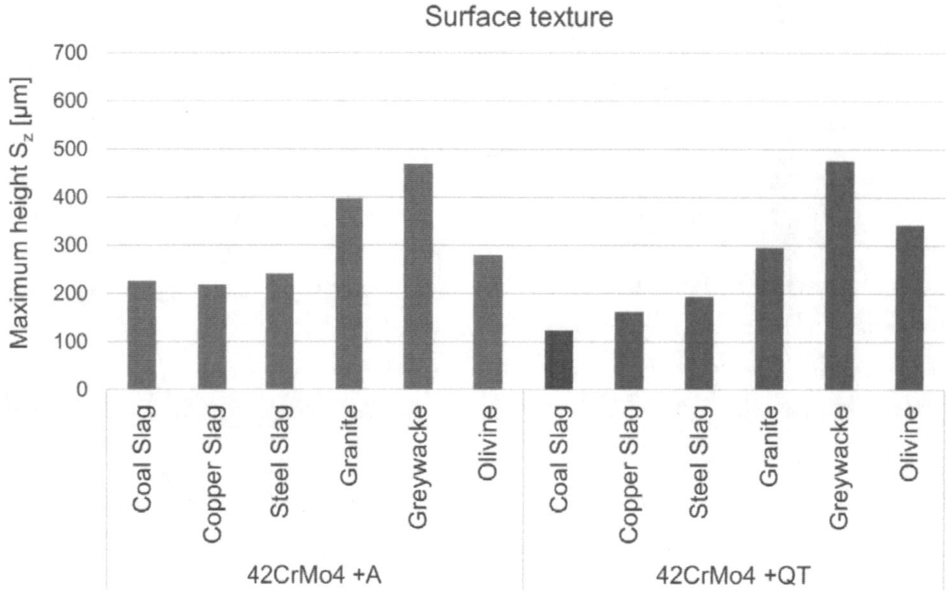

Figure 8: Controlled-depth machining (CDM) with alternative abrasives. Surface texture by the maximum height S_z on a heat-treated 42CrMo4 steel. The results on the annealed 42CrMo4 +A are on the left side, the quenched and tempered 42CrMo4 +QT is on the right.

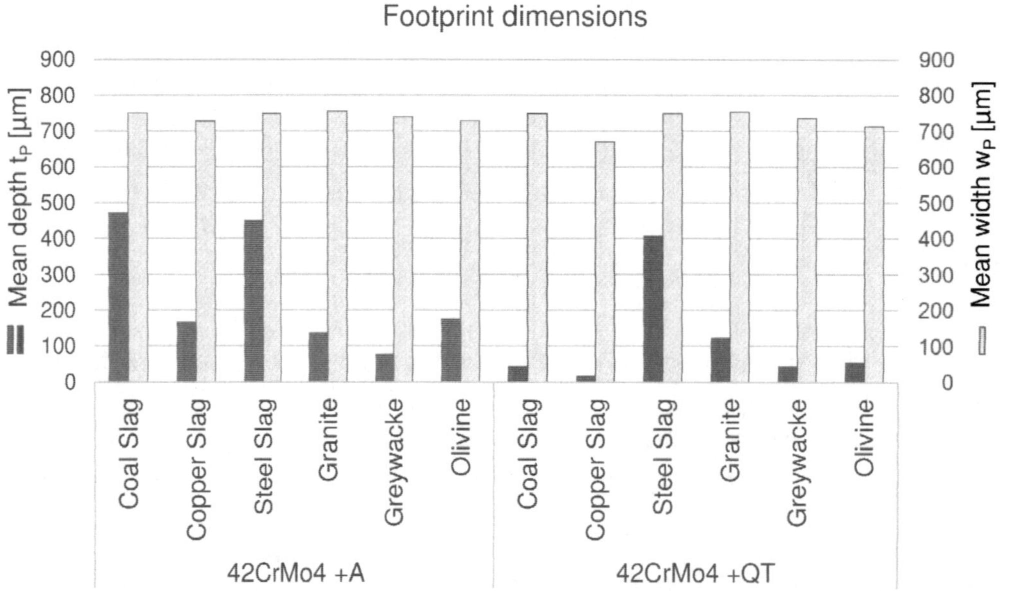

Figure 9: Controlled-depth machining (CDM) with alternative abrasives. Footprint dimensions by the mean depth t_P and the mean width w_P on a heat-treated 42CrMo4 steel. The results on the annealed 42CrMo4 +A are on the left side, the quenched and tempered 42CrMo4 +QT is on the right.

In the interim conclusion for CDM, the assessment of the findings stood out at some points. Firstly, the garnet abrasives behaved in a fundamentally similar way to CT, although the contrasts were more pronounced at CDM. Consequently, the optimization potential for certain applications became even more obvious. Although the high MRR of the garnet abrasives could not be achieved here, the considered alternatives nevertheless convinced with better surface texture. The results revealed a high potential of using alternative abrasives in AWJ machining. However, an inferior performance on the experimental conditions of this study does not necessarily mean a final rating of alternative abrasives. For example, a less performing abrasive here could be competitive with other materials, for example when cutting fibre-reinforced plastics or non-ferrous metals.

4. SUMMARY AND OUTLOOK

In this study, a fundamental benchmark on different abrasives was conducted experimentally. The aim was to rate different sources of natural and synthetic minerals for a resource-efficient approach in abrasive waterjet (AWJ) technology, here by machining a 42CrMo4 steel alloy of two structural modifications by heat treatment. By either ductile or brittle properties of the workpiece material, a comparison was made between cutting through (CT) the entire workpiece and controlled-depth machining (CDM), often referred to as waterjet milling.

The evaluation of this study highlighted the optimization potential when using different abrasives in various applications of AWJ. For a practical benchmark of abrasives, a reference was proposed in comparison with an average of the typical garnets. The cutting performance in CT with garnets was very similar and on par with MRR in CDM on the considered material. However, it was shown that alternative abrasives showed great potential towards a stable, high-quality process for CDM strategies or could act as an add-on to garnets to gain economic application. At the same time, by including waste products from other industries, a contribution to resource-savings in AWJ machining was made.

In the future, this study will be extended toward the reusability of abrasives. To benchmark the closed-loop capability, the systematic distribution of particle sizes [17] and the abrasive wear after workpiece contact must be considered. The abrasives should also be examined to hold the permissible limits for discharging wastewater. To be on time with upcoming regulations, the corresponding treatment of wastewater should be integrated into waterjet machines on a manufacturer level. Further experiments should be conducted to characterize the machining result, e.g. by the constriction of the engaged AWJ tool or by surface integrity aspects. The approach of this study to rate against a garnet reference could serve as a basis to identify alternative abrasives and benchmark them on other materials.

5. ACKNOWLEDGMENTS

The work was funded within the KMU-innovativ project (program for SMEs) "Resource-saving and Economical Production through Abrasive Waterjet Technology (ReWiProAWS)" by the German Ministry of Education and Research under the supervision of the Project Management Agency Karlsruhe (PTKA).

6. REFERENCES

[1] Klocke, F., Schmitt, R., Zeis, M., Heidemanns, L. et al., 2015. Technological and Economical Assessment of Alternative Process Chains for Blisk Manufacture. Procedia CIRP 35.

[2] Moreno, J.R.S., Ribeiro, A.A., Souza Gonçalves, J.F. de, 2021. Comparative Analysis on the Wear Performance of New and Reconditioned Ceramic Cutting Tools When Machining Nodular Cast Iron GGG-60, with Lubrication of Two Types of Fluids. Journal of Materials Engineering and Performance Vol. 30.

[3] Brecher, C., Krüger, J., Uhlmann, E., Verl, A. et al., 2012. Fortschritt Berichte VDI, Ressourceneffiziente Produktion.

[4] Axinte, D.A., Karpuschewski, B., Kong, M.C., Beaucamp, A.T. et al., 2014. High Energy Fluid Jet Machining (HEFJet-Mach): From scientific and technological advances to niche industrial applications, CIRP Annals – Manufacturing Technology 63.

[5] Henning, A., Miles, P., Stang, D., 2011. Efficient Operation of Abrasive Waterjet Cutting in industrial applications. 2011 WJTA-IMCA Conference and Expo.

[6] Fowler Gary, 2003. Abrasive Water-jet - Controlled Depth Milling of Titanium Alloys. PHD Thesis, University of Nottingham.

[7] Hashish, M., Editor, 1987. MILLING WITH ABRASIVE-WATERJETS: A PRELIMINARY INVESTIGATION. PROCEEDINGS OF THE FOURTH U.S. WATER JET CONFERENCE.

[8] Momber, A.W., Kovacevic, R., 1997. Princples of Abrasive Water Jet Machining. Springer, Berlin Heidelberg New York.

[9] Schüler, M., Heidrich, D., Herrig, T., Fang, X.F. et al., 2021. Automotive hybrid design production and effective end machining by novel abrasive waterjet technique. Procedia CIRP 101.

[10] Schüler, M., Dadgar, M., Herrig, T., Klink, A. et al., 2021. Experimental investigation of abrasive properties in waterjet machining. Procedia CIRP 101.

[11] Voynick, S. Garnet: An Industrial Commodity. Online article. https://www.rockngem.com/garnet-an-industrial-commodity/.

[12] Briggs, J. The Abrasives Industry in Europe and North America. Online article. https://www.glassonweb.com/news/abrasive-industry-europe-and-north-america.

[13] Kantha Babu, M., Krishnaiah Chetty, O., 2003. A study on recycling of abrasives in abrasive water jet machining. Wear 254.

[14] Perec, A., 2018. Environmental Aspects of Abrasive Water Jet Cutting. Annual Set The Environment Protection, Vol. 20.

[15] Kurd, M.O., 2004. The Material and Energy Flow Through the Abrasive Waterjet Machining and Recycling Processes. Thesis, Massachusetts Institute of Technology.

[16] Pi, V.N., Hung, T.Q., 2018. Recycling and Recharging of Supreme Garnet in Abrasive Waterjet Machining, in Abrasive Technology - Characteristics and Applications, InTechOpen.

[17] Bergs, T., Schüler, M., Herrig, T., Fernolendt, J. et al., 2019. Systematic Change of Abrasive Size Distribution. Advances in Water Jetting. Springer.

[18] Alsoufi, M.S., 2017. State-of-the-Art in Abrasive Water Jet Cutting Technology and the Promise for Micro- and Nano-Machining. International Journal of Mechanical Engineering and Applications.
[19] Fowler G, P.H. Shipway∗, I.R. Pashby, 2004. A technical note on grit embedment following abrasive water-jet milling of a titanium alloy. Journal of Materials Processing Technology 159.
[20] MieSzala, M., Torrubia, P.L., Axinte, D.A., Schwiedrzik, J.J. et al., 2017. Erosion mechanisms during abrasive waterjet machining: Model microstructures and single particle experiments. Journal of Materials Processing Tech 247.

7. NOMENCLATURE

AWJ	Abrasive Water Jet
CT	Cutting through
CDM	Controlled-depth machining
HRC	Hardness Rockwell
$\emptyset \boldsymbol{d_O}$	Orifice inner diameter in mm
$\emptyset \boldsymbol{d_W}$	Focus tube inner diameter in mm
SoD	Stand-off distance in mm
$\dot{m}_A$	Abrasive mass flow in g/min
p	Water pressure in MPa
v_F	Feedrate in mm/min
t_C	Derived cutting thickness in mm
MRR	Material removal rate in mm^3/min
S_z	Maximum surface height in µm
w_P	Width of the profile in µm
d_P	Depth of the profile µm
$\overline{t_C}$	Derived cutting thickness in mm
$\overline{MRR}$	Average material removal rate of garnets as reference in mm^3/min
$\overline{S_Z}$	Average maximum surface height of garnets as reference in µm
$\overline{w_P}$	Average width of the profile of garnets as reference in µm
$\overline{d_P}$	Average depth of the profile of garnets as reference µm

AUTHOR INDEX